Gaia

A new look at life on Earth

J. E. Lovelock

Gaia

A new look at life on Earth

Oxford New York Toronto Melbourne
OXFORD UNIVERSITY PRESS
1979

Oxford University Press, Walton Street, Oxford OX2 6DP

OXFORD LONDON GLASGOW
NEW YORK TORONTO MELBOURNE WELLINGTON
KUALA LUMPUR SINGAPORE JAKARTA HONG KONG TOKYO
DELHI BOMBAY CALCUTTA MADRAS KARACHI
IBADAN NAIROBI DAR ES SALAAM CAPE TOWN

British Library Cataloguing in Publication Data
Lovelock, J. E.
  Gaia.
  1. Ecology
  I. Title
  500.9   QH541   79-40499

  ISBN 0-19-217665-X

*Printed in Great Britain by
The Pitman Press, Bath*

Contents

Preface

The concept of Mother Earth or, as the Greeks called her long ago, Gaia, has been widely held throughout history and has been the basis of a belief which still coexists with the great religions. As a result of the accumulation of evidence about the natural environment and the growth of the science of ecology, there have recently been speculations that the biosphere may be more than just the complete range of all living things within their natural habitat of soil, sea, and air. Ancient belief and modern knowledge have fused emotionally in the awe with which astronauts with their own eyes and we by indirect vision have seen the Earth revealed in all its shining beauty against the deep darkness of space. Yet this feeling, however strong, does not prove that Mother Earth lives. Like a religious belief, it is scientifically untestable and therefore incapable in its own context of further rationalization.

Journeys into space did more than present the Earth in a new perspective. They also sent back information about its atmosphere and its surface which provided a new insight into the interactions between the living and the inorganic parts of the planet. From this has arisen the hypothesis, the model, in which the Earth's living matter, air, oceans, and land surface form a complex system which can be seen as a single organism and which has the capacity to keep our planet a fit place for life.

This book is a personal account of a journey through space and time in search of evidence with which to substantiate this model of the Earth. The quest began about fifteen years ago and has ranged through the territories of many different scientific disciplines, indeed from astronomy to zoology.

Such journeys are lively, for the boundaries between the sciences are jealously guarded by their Professors and within each territory there is a different arcane language to be learnt. In the ordinary way a grand tour of this kind would be extravagantly expensive and unproductive in its yield of new knowledge; but just as trade often still goes on between nations at war, it is also possible for a chemist to travel through such distant disciplines as meteorology or physiology, if he has something to barter. Usually this is a piece of hardware or a technique. I was fortunate to work briefly with A. J. P. Martin, who developed among other things the important chemical analytical technique of gas chromatography. During that time I added some embellishments which extended the range of his invention. One of these was the so-called electron capture detector. This device is notable for its exquisite sensitivity in the detection of traces of certain chemical substances. This sensitivity first made possible the discovery that pesticide residues were present in all creatures of the Earth, from penguins in Antarctica to the milk of nursing mothers in the USA. It was this discovery that facilitated the writing of Rachel Carson's immensely influential book, *Silent Spring*, by providing the evidence needed to justify her concern over the damage done to the biosphere by the ubiquitous presence of these toxic chemicals. The electron capture detector has continued to reveal minute but significant quantities of other toxic chemicals in places where they ought not to be. Among these intruders are: PAN (peroxyacetyl nitrate), a toxic component of Los Angeles smog; the PCBs (polychlorobiphenyls) in the remote natural environment; and most recently, in the atmosphere at large, the chlorofluorocarbons and nitrous oxide—substances which are thought to deplete the strength of ozone in the stratosphere.

Electron capture detectors were undoubtedly the most valued of the trade goods which enabled me to pursue my quest for Gaia through the various scientific disciplines, and also indeed to travel in literal fashion around the Earth itself. However, although my role as a tradesman made interdisciplinary journeys feasible, they have not been easy, since the past fifteen years have witnessed a great deal of turmoil in the life sciences, particularly in areas where science has been drawn into the processes of power politics.

When Rachel Carson made us aware of the dangers arising from the mass application of toxic chemicals, she presented her arguments in the manner of an advocate rather than that of a scientist. In other words, she selected the evidence to prove her case. The chemical industry, seeing its livelihood threatened by her action, responded with an equally selective set of arguments, chosen in defence. This may have been a fine way of achieving justice for the people in those aspects of the problem affecting the community at large, and perhaps in this instance it was scientifically excusable; but it seems to have established a pattern. Since then, a great deal of scientific argument and evidence concerning the environment is presented as if in a court-room or at a public inquiry. It cannot be said too often that, although this may be good for the democratic process of public participation in matters of general concern, it is not the best way to discover scientific truth. Truth is said to be the first casualty of war. It is also weakened by being used selectively in evidence to prove a case in law.

On environmental matters the scientific community seems to be divided into collectivized warring groups, and there are strong pressures to conform to the dogma of whatever tribe one happens to be with. The first six chapters of this book are not concerned with matters of social controversy—at least not yet. In the last three chapters, however, which are about Gaia and mankind, I am aware of having moved on to a battleground where powerful forces are in action.

Sir Alan Parkes, in his book *Sex, Science and Society*, commented that 'Science can be serious without being sacrosanct.' I have tried to keep these wise words in mind throughout, but sometimes the task of writing for a general readership on matters normally expressed in precise but esoteric language has all but defeated me. In consequence there are passages and sentences which may read as if infected with the twin blights of anthropomorphism and teleology.

I have frequently used the word Gaia as a shorthand for the hypothesis itself, namely that the biosphere is a self-regulating entity with the capacity to keep our planet healthy by controlling the chemical and physical environment. Occasionally it has been difficult, without excessive circumlocution, to avoid talking of Gaia as if she were known to be sentient. This is meant no more seriously than is the appellation 'she' when given to a ship by those who sail in her, as a recognition that even pieces of wood and metal when specifically designed and assembled may achieve a composite identity with its own characteristic signature, as distinct from being the mere sum of its parts.

Shortly after writing this book I came across an article by Alfred Redfield in the *American Scientist* of 1958. In it he put forward the hypothesis that the chemical composition of the atmosphere and of the oceans was biologically controlled. He produced supporting evidence from the distribution of the elements. I am glad that I saw Redfield's contribution to the development of the Gaia hypothesis in time to acknowledge it here, but I realize that there must have been many others who had these and similar thoughts and some may have published them. The notion of Gaia, of a living Earth, has not in the past been acceptable to the main science stream and consequently seeds sown in earlier times would not have flourished, but instead would have remained buried in the deep mulch of scientific papers.

In a subject so broadly based as that of this book there was the need for much advice and I wish to thank the many scientific colleagues who patiently and unstintingly gave their time to help me, especially Professor Lynn Margulis of Boston who has been my constant colleague and guide. I am also grateful to Professor C. E. Junge of Mainz and to Professor B. Bollin of Stockholm, who first encouraged me to write about Gaia; and to my colleagues Dr James Lodge of Boulder, Colorado, Sidney Epton of Shell Research Ltd., and Peter Fellgett of Reading, who encouraged me to continue the quest.

My special thanks go to Evelyn Frazer, who took the draft of this book and turned the disordered mosaic of sentences and paragraphs into a readable whole, and did it so skilfully that the result conveys what I want to say in the way I would have said it if I could.

Finally, I wish to record my debt to Helen Lovelock, who not only produced the typescript but also established and kept the environment in which writing and thinking were possible.

I have listed at the end of the book, arranged by chapter, the principal sources of information and suggestions for additional reading, together with some definitions and explanations of terms and of the system of units and measurements used in the text.

CHAPTER 1

Introductory

As I write, two Viking spacecraft are circling our fellow planet Mars, awaiting landfall instructions from the Earth. Their mission is to search for life, or evidence of life, now or long ago. This book also is about a search for life, and the quest for Gaia is an attempt to find the largest living creature on Earth. Our journey may reveal no more than the almost infinite variety of living forms which have proliferated over the Earth's surface under the transparent case of the air and which constitute the biosphere. But if Gaia does exist, then we may find ourselves and all other living things to be parts and partners of a vast being who in her entirety has the power to maintain our planet as a fit and comfortable habitat for life.

The quest for Gaia began more than fifteen years ago, when NASA (the National Aeronautics and Space Administration of the USA) first made plans to look for life on Mars. It is therefore right and proper that this book should open with a tribute to the fantastic Martian voyage of those two mechanical Norsemen.

In the early nineteen-sixties I often visited the Jet Propulsion Laboratories of the California Institute of Technology in Pasadena, as consultant to a team, later to be led by that most able of space biologists Norman Horowitz, whose main objective was to devise ways and means of detecting life on Mars and other planets. Although my particular brief was to advise on some comparatively simple problems of instrument design, as one whose childhood was illuminated by the writings of Jules Verne and Olaf Stapledon I was delighted to have the chance of discussing at first hand the plans for investigating Mars.

At that time, the planning of experiments was mostly based on the assumption that evidence for life on Mars would be much the same as for life on Earth. Thus one proposed series of experiments involved dispatching what was, in effect, an automated microbiological laboratory to sample the Martian soil and judge its suitability to support bacteria, fungi, or other micro-organisms. Additional soil experiments were designed to test for chemicals whose presence would indicate life at work: proteins, amino-acids, and particularly optically active substances with the capacity that organic matter has to twist a beam of polarized light in a counter-clockwise direction.

After a year or so, and perhaps because I was not directly involved, the euphoria arising from my association with this enthralling problem began to subside, and I found myself asking some rather down-to-earth questions, such as, 'How can we be sure that the Martian way of life, if any, will reveal itself to tests based on Earth's life style?' To say nothing of more difficult questions, such as, 'What is life, and how should it be recognized?'

Some of my still sanguine colleagues at the Jet Propulsion Laboratories mistook my growing scepticism for cynical disillusion and quite properly asked, 'Well, what would you do instead?' At that time I could only reply vaguely, 'I'd look for an entropy reduction, since this must be a general characteristic of all forms of life.' Understandably, this reply was taken to be at the best unpractical and at worst plain obfuscation, for few physical concepts can have caused as much confusion and misunderstanding as has that of entropy.

It is almost a synonym for disorder and yet, as a measure of the rate of dissipation of a system's thermal energy, it can be precisely expressed in mathematical terms. It has been the bane of generations of students and is direfully associated in many minds with decline and decay, since its expression in the Second Law of Thermodynamics (indicating that all energy will eventually dissipate into heat universally distributed and will no longer be available for the performance of useful work) implies the predestined and inevitable run-down and death of the Universe.

Although my tentative suggestion had been rejected, the idea of looking for a reduction or reversal of entropy as a sign of life had implanted itself in my mind. It grew and waxed fruitful until, with the help of many colleagues, Dian Hitchcock, Sidney Epton, Peter Simmonds, and especially Lynn Margulis, it evolved into the hypothesis which is the subject of this book.

Back home in the quiet countryside of Wiltshire, after my visits to the Jet Propulsion Laboratories, I had time to do more thinking and reading about the real character of life and how one might recognize it anywhere and in any guise. I expected to discover somewhere in the scientific literature a comprehensive definition of life as a physical process, on which one could base the design of life-detection experiments, but I was surprised to find how little had been written about the nature of life itself. The present interest in ecology and the application of systems analysis to biology had barely begun and there was still in those days the dusty academic air of the classroom about the life sciences. Data galore had been accumulated on every conceivable aspect of living species, from their outermost to their innermost parts, but in the whole vast encyclopaedia of facts the crux of the matter, life itself, was almost totally ignored. At best, the literature read like a collection of expert reports, as if a group of scientists from another world had taken a television receiver home with them and had reported on it. The chemist said it was made of wood, glass, and metal. The physicist said it radiated heat and light. The engineer said the supporting wheels were too small and in the wrong place for it to run smoothly on a flat surface. But nobody said what it was.

This seeming conspiracy of silence may have been due in part to the division of science into separate disciplines, with each specialist assuming that someone else has done the job. Some biologists may believe that the process of life is adequately described by some mathematical theorem of physics or cybernetics, and some physicists may assume that it is factually described in the recondite writings on molecular biology which one day he will find time to read. But the most probable cause of our closed minds on the subject is that we already have a very rapid, highly efficient life-recognition programme in our inherited set of instincts, our 'read-only' memory as it might be called in computer technology. Our recognition of living things, both animal and vegetable, is instant and automatic, and our fellow-creatures in the animal world appear to have the same facility. This powerful and effective but unconscious process of recognition no doubt originally evolved as a survival factor. Anything living may be edible, lethal, friendly, aggressive, or a potential mate, all questions of prime significance for our welfare and continued existence. However, our automatic recognition system appears to have paralysed our capacity for conscious thought about a definition of life. For why should we need to define what is obvious and unmistakable in all its manifestations, thanks to our built-in programme? Perhaps for that very reason, it is an automatic process operating without conscious understanding, like the autopilot of an aircraft.

Even the new science of cybernetics has not tackled the problem, although it is concerned with the mode of operation of all manner of systems from the simplicity of a valve-operated water tank to the complex visual control process which enables your eyes to scan this page. Much, indeed, has already been said and written about the cybernetics of artificial intelligence, but the question of defining real life in cybernetic terms remains unanswered and is seldom discussed.

During the present century a few physicists have tried to define life. Bernal, Schroedinger, and Wigner all came to the same general conclusion, that life is a member of the class of phenomena which are open or continuous systems able to decrease their internal entropy at the expense of substances or free energy taken in from the environment and subsequently rejected in a degraded form. This definition is not only difficult to grasp but is far too general to apply to the specific detection of life. A rough paraphrase might be that life is one of those processes which are found whenever there is an abundant flow of energy. It is characterized by a tendency to shape or form itself as it consumes, but to do so it must always excrete low-grade products to the surroundings.

We can now see that this definition would apply equally well to eddies in a flowing stream, to hurricanes, to flames, or even to refrigerators and many other man-made contrivances. A flame assumes a characteristic shape as it burns, and needs an adequate supply of fuel and air to keep going, and we are now only too well aware that the pleasant warmth and dancing flames of an open fire have to be paid for in the excretion of waste heat and pollutant gases. Entropy is reduced locally by the flame formation, but the overall total of entropy is increased during the fuel consumption.

Yet even if too broad and vague, this classification of life at least points us in the right direction. It suggests, for example, that there is a boundary, or interface, between the 'factory' area where the flow of energy or raw materials is put to work and entropy is consequently reduced, and the surrounding environment which receives the discarded waste products. It also suggests that life-like processes require a flux of energy above some minimal value in order to get going and keep going. The nineteenth-century physicist Reynolds observed that turbulent eddies in gases and liquids could only form if the rate of flow was above some critical value in relation to local conditions. The Reynolds dimensionless number can be calculated from simple knowledge of a fluid's properties and its local flow boundaries. Similarly, for life to begin, not only the quantity but also the quality, or potential, of the energy flow must be sufficient. If, for example, the sun's surface temperature were 500 degrees instead of 5,000 degrees Centigrade and the Earth were correspondingly closer, so that we received the same amount of warmth, there would be little difference in climate, but life would never have got going. Life needs energy potent enough to sever chemical bonds; mere warmth is not enough.

It might be a step forward if we could establish dimensionless numbers like the Reynolds scale to characterize the energy conditions of a planet. Then those enjoying, with the Earth, a flux of free solar energy above these critical values would predictably have life whilst those low on the scale, like the cold outer planets, would not.

The design of a universal life-detection experiment based on entropy reduction seemed at this time to be a somewhat unpromising exercise. However, assuming that life on any planet would be bound to use the fluid media—oceans, atmosphere, or both—as conveyor-belts for raw materials and waste products, it occurred to me that some of the activity associated with concentrated entropy reduction within a living system might spill over into the conveyor-belt regions and alter their composition. The atmosphere of a life-bearing planet would thus become recognizably different from that of a dead planet.

Mars has no oceans. If life had established itself there, it would have had to make use of the atmosphere or stagnate. Mars therefore seemed a suitable planet for a life-detection exercise based on chemical analysis of the atmosphere. Moreover, this could be carried out regardless of the choice of landing site. Most life-detection experiments are effective only within a suitable target area. Even on Earth, local search techniques would be unlikely to yield much positive evidence of life if the landfall occurred on the Antarctic ice sheet or the Sahara desert or in the middle of a salt lake.

While I was thinking on these lines, Dian Hitchcock visited the Jet Propulsion Laboratories. Her task was to compare and evaluate the logic and information-potential of the many suggestions for detecting life on Mars. The notion of life detection by atmospheric analysis appealed to her, and we began developing the idea together. Using our own planet as a model, we examined the extent to which simple knowledge of the chemical composition of the Earth's atmosphere, when coupled with such readily accessible information as the degree of solar radiation and the presence of oceans as well as land masses on the Earth's surface, could provide evidence for life.

Our results convinced us that the only feasible explanation of the Earth's highly improbable atmosphere was that it was being manipulated on a day-to-day basis from the surface, and that the manipulator was life itself. The significant decrease in entropy—or, as a chemist would put it, the persistent state of disequilibrium among the atmospheric gases—was on its own clear proof of life's activity. Take, for example, the simultaneous presence of methane and oxygen in our atmosphere. In sunlight, these two gases react chemically to give carbon dioxide and water vapour. The rate of this reaction is such that to sustain the amount of methane always present in the air, at least 1,000 million tons of this gas must be introduced into the atmosphere yearly. In addition, there must be some means of replacing the oxygen used up in oxidizing methane and this requires a production of at least twice as much oxygen as methane. The quantities of both of these gases required to keep the Earth's extraordinary atmospheric mixture constant was improbable on an abiological basis by at least 100 orders of magnitude.

Here, in one comparatively simple test, was convincing evidence for life on Earth, evidence moreover which could be picked up by an infra-red telescope sited as far away as Mars. The same argument applies to other atmospheric gases, especially to the ensemble of reactive gases constituting the atmosphere as a whole. The presence of nitrous oxide and of ammonia is as anomalous as that of methane in our oxidizing atmosphere. Even nitrogen in gaseous form is out of place, for with the Earth's abundant and neutral oceans, we should expect to find this element in the chemically stable form of the nitrate ion dissolved in the sea.

Our findings and conclusions were, of course, very much out of step with conventional geochemical wisdom in the mid-sixties. With some exceptions, notably Rubey, Hutchinson, Bates, and Nicolet, most geochemists regarded the atmosphere as an end-product of planetary outgassing and held that subsequent reactions by abiological processes had determined its present state. Oxygen, for example, was thought to come solely from the breakdown of water vapour and the escape of hydrogen into space, leaving an excess of oxygen behind. Life merely borrowed gases from the atmosphere and returned them unchanged. Our contrasting view required an atmosphere which was a dynamic extension of the biosphere itself. It was not easy to find a journal prepared to publish so radical a notion but, after several rejections, we found an editor, Carl Sagan, prepared to publish it in his journal, *Icarus*.

Nevertheless, considered solely as a life-detection experiment, atmospheric analysis was, if anything, too successful. Even then, enough was known about the Martian atmosphere to suggest that it consisted mostly of carbon dioxide and showed no signs of the exotic chemistry characteristic of Earth's atmosphere. The implication that Mars was probably a lifeless planet was unwelcome news to our sponsors in space research. To make matters worse, in September 1965 the US

Congress decided to abandon the first Martian exploration programme, then called Voyager. For the next year or so, ideas about looking for life on other planets were to be discouraged.

Space exploration has always served as a convenient whipping-boy to those needing money for some worthy cause, yet it is far less expensive than many a stuck-in-the-mud, down-to-earth technological failure. Unfortunately, the apologists for space science always seem over-impressed by engineering trivia and make far too much of non-stick frying pans and perfect ball-bearings. To my mind, the outstanding spin-off from space research is not new technology. The real bonus has been that for the first time in human history we have had a chance to look at the Earth from space, and the information gained from seeing from the outside our azure-green planet in all its global beauty has given rise to a whole new set of questions and answers. Similarly, thinking about life on Mars gave some of us a fresh standpoint from which to consider life on Earth and led us to formulate a new, or perhaps revive a very ancient, concept of the relationship between the Earth and its biosphere.

By great good fortune, so far as I was concerned, the nadir of the space programme coincided with an invitation from Shell Research Limited for me to consider the possible global consequences of air pollution from such causes as the ever-increasing rate of combustion of fossil fuels. This was in 1966, three years before the formation of Friends of the Earth and similar pressure-groups brought pollution problems to the forefront of the public mind.

Like artists, independent scientists need sponsors but this rarely involves a possessive relationship. Freedom of thought is the rule. This should hardly need saying, but nowadays many otherwise intelligent individuals are conditioned to believe that all research work supported by a multi-national corporation must be suspect by origin. Others are just as convinced that similar work coming from an institution in a Communist country will have been subject to Marxist theoretical constraint and will therefore be diminished. The ideas and opinions expressed in this book are inevitably influenced to some degree by the society in which I live and work, and especially by close contact with numerous scientific colleagues in the West. So far as I know, these mild pressures are the only ones which have been exerted on me.

The link between my involvement in problems of global air pollution and my previous work on life detection by atmospheric analysis was, of course, the idea that the atmosphere might be an extension of the biosphere. It seemed to me that any attempt to understand the consequences of air pollution would be incomplete and probably ineffectual if the possibility of a response or an adaptation by the biosphere was overlooked. The effects of poison on a man are greatly modified by his capacity to metabolize or excrete it; and the effect of loading a biospherically controlled atmosphere with the products of fossil fuel combustion might be very different from the effect on a passive inorganic atmosphere. Adaptive changes might take place which would lessen the perturbations due, for instance, to the accumulation of carbon dioxide. Or the perturbations might trigger some compensatory change, perhaps in the climate, which would be good for the biosphere as a whole but bad for man as a species.

Working in a new intellectual environment, I was able to forget Mars and to concentrate on the Earth and the nature of its atmosphere. The result of this more single-minded approach was the development of the hypothesis that the entire range of living matter on Earth, from whales to viruses, and from oaks to algae, could be regarded as constituting a single living entity, capable of manipulating the Earth's atmosphere to suit its overall needs and endowed with faculties and powers far beyond those of its constituent parts.

It is a long way from a plausible life-detection experiment to the hypothesis that the Earth's atmosphere is actively maintained and regulated by life on the surface, that is, by the biosphere. Much of this book deals with more recent evidence in support of this view. In 1967 the reasons for making the hypothetical stride were briefly these:

· Life first appeared on the Earth about 3,500 million years ago. From that time until now, the presence of fossils shows that the Earth's climate has changed very little. Yet the output of heat from the sun, the surface properties of the Earth, and the composition of the atmosphere have almost certainly varied greatly over the same period.

The chemical composition of the atmosphere bears no relation to the expectations of steady-state chemical equilibrium. The presence of methane, nitrous oxide, and even nitrogen in our present oxidizing atmosphere represents violation of the rules of chemistry to be measured in tens of orders of magnitude. Disequilibria on this scale suggest that the atmosphere is not merely a biological product, but more probably a biological construction: not living, but like a cat's fur, a bird's feathers, or the paper of a wasp's nest, an extension of a living system designed to maintain a chosen environment. Thus the atmospheric concentration of gases such as oxygen and ammonia is found to be kept at an optimum value from which even small departures could have disastrous consequences for life.

The climate and the chemical properties of the Earth now and throughout its history seem always to have been optimal for life. For this to have happened by chance is as unlikely as to survive unscathed a drive blindfold through rush-hour traffic.

By now a planet-sized entity, albeit hypothetical, had been born, with properties which could not be predicted from the sum of its parts. It needed a name. Fortunately the author William Golding was a fellow-villager. Without hesitation he recommended that this creature be called Gaia, after the Greek Earth goddess also known as Ge, from which root the sciences of geography and geology derive their names. In spite of my ignorance of the classics, the suitability of this choice was obvious. It was a real four-lettered word and would thus forestall the creation of barbarous acronyms, such as Biocybernetic Universal System Tendency/Homoeostasis. I felt also that in the days of Ancient Greece the concept itself was probably a familiar aspect of life, even if not formally expressed. Scientists are usually condemned to lead urban lives, but I find that country people still living close to the earth often seem puzzled that anyone should need to make a

formal proposition of anything as obvious as the Gaia hypothesis. For them it is true and always has been.

I first put forward the Gaia hypothesis at a scientific
meeting about the origins of life on Earth which took place in
Princeton, New Jersey, in 1969. Perhaps it was poorly presented. It certainly did not appeal to anyone except Lars
Gunnar Sillen, the Swedish chemist now sadly dead, and Lynn
Margulis, of Boston University, who had the task of editing our
various contributions. A year later in Boston Lynn and I met
again and began a most rewarding collaboration which, with
her deep knowledge and insight as a life scientist, was to go far
in adding substance to the wraith of Gaia, and which still
happily continues.

We have since defined Gaia as a complex entity involving
the Earth's biosphere, atmosphere, oceans, and soil; the total-
ity constituting a feedback or cybernetic system which seeks
an optimal physical and chemical environment for life on this
planet. The maintenance of relatively constant conditions by
active control may be conveniently described by the term
'homoeostasis'.

Gaia has remained a hypothesis but, like other useful
hypotheses, she has already proved her theoretical value, if not
her existence, by giving rise to experimental questions and
answers which were profitable exercises in themselves. If, for
example, the atmosphere is, among other things, a device for
conveying raw materials to and from the biosphere, it would be
reasonable to assume the presence of carrier compounds for
elements essential in all biological systems, for example,
iodine and sulphur. It was rewarding to find evidence that both
were conveyed from the oceans, where they are abundant,
through the air to the land surface, where they are in short
supply. The carrier compounds, methyl iodide and dimethyl
sulphide respectively, are directly produced by marine life.
Scientific curiosity being unquenchable, the presence of these
interesting compounds in the atmosphere would no doubt have
been discovered in the end and their importance discussed
without the stimulus of the Gaia hypothesis. But they were
actively sought as a result of the hypothesis and their presence
was consistent with it.

If Gaia exists, the relationship between her and man, a dominant animal species in the complex living system, and the possibly shifting balance of power between them, are questions of obvious importance. I have discussed them in later chapters, but this book is written primarily to stimulate and entertain. The Gaia hypothesis is for those who like to walk or simply stand and stare, to wonder about the Earth and the life it bears, and to speculate about the consequences of our own presence here. It is an alternative to that pessimistic view which sees nature as a primitive force to be subdued and conquered. It is also an alternative to that equally depressing picture of our planet as a demented spaceship, forever travelling, driverless and purposeless, around an inner circle of the sun.

CHAPTER 2

In the beginning

In scientific usage, an aeon represents 1,000 million years. So far as we can tell from the record of the rocks and from measurements of their radioactivity, the Earth began its existence as a separate body in space about 4,500 million years, or four and a half aeons, ago. The earliest traces of life so far identified are to be found in sedimentary rocks formed more than three aeons ago. However, as H. G. Wells put it, the record of the rocks is no more a complete record of life in the past than the books of a bank are a record of the existence of everybody in the neighbourhood. Untold millions of early life-forms and their more complex but still soft-bodied descendants may have lived and flourished and passed away without putting anything by for the future, or—to change the simile—without leaving any traces, let alone skeletons for the geological cupboard.

It is not surprising, therefore, that little is known about the origin of life on our planet and still less about the course of its early evolution. But if we review what we know concerning the Earth's beginnings in the context of the universe from which it was formed, we can at least make intelligent guesses about the environment in which life, and potentially Gaia, began, and set about ensuring their mutual survival.

We know, from observations of events in our own galaxy, that the stellar universe resembles a living population, in which at any time may be found people of all ages from infants to centenarians. As old stars, like old soldiers, fade away, while others expire more spectacularly in an explosive blaze of glory, fresh incandescent globes with their satellite moths are taking shape. When we examine spectroscopically the interstellar dust and gas clouds from which new suns and planets condense, we find that these contain an abundance of the simple and compound molecules from which the chemical building blocks of life can be assembled. Indeed, the universe appears to be littered with life's chemicals. Nearly every week there is news from the astronomical front of yet another complex organic substance found far away in space. It seems almost as if our galaxy were a giant warehouse containing the spare parts needed for life.

If we can imagine a planet made of nothing but the component parts of watches, we may reasonably assume that in the fullness of time—perhaps 1,000 million years—gravitational forces and the restless motion of the wind would assemble at least one working watch. Life on Earth probably started in a similar manner. The countless number and variety of random encounters between individual molecular components of life may have eventually resulted in a chance association of parts which together could perform a life-like task, such as gathering sunlight and using its energy to contrive some further action which would otherwise have been impossible or forbidden by the laws of physics. (The ancient Greek myth of Prometheus stealing fire from heaven and the biblical story of Adam and Eve tasting the forbidden fruit may have far deeper roots in our ancestral history than we realize.) Later, as more of these primitive assembly-forms appeared, some successfully combined and from their union more complex assemblies emerged with new properties and powers, and united in their turn, the product of fruitful associations being always a more potent assembly of working parts, until eventually there came into being a complex entity with the properties of life itself: the first micro-organism and one capable of using sunlight and the molecules of the environment to produce its own duplicate.

The odds against such a sequence of encounters leading to the first living entity are enormous. On the other hand, the number of random encounters between the component molecules of the Earth's primaeval substance must have been incalculable. Life was thus an almost utterly improbable event with almost infinite opportunities of happening. So it did. Let us at least assume that it happened in this way rather than by the mysterious planting of a seed, or the drift of spores from elsewhere or indeed by outside intervention of any kind. We are not primarily concerned with the origin of life but with the relationship between the evolving biosphere and the early planetary environment of the Earth.

What was the state of the Earth just before life began, perhaps three and a half aeons ago? Why was our planet able to bear and sustain life when its nearest siblings, Mars and Venus, apparently failed? What hazards and near disasters would have faced the infant biosphere and how might Gaia's presence have helped to surmount them? To suggest possible answers to these intriguing questions, we must first return to the circumstances in which the Earth itself was formed, some four and a half aeons ago.

It seems almost certain that close in time and space to the origin of our solar system, there was a supernova event. A supernova is the explosion of a large star. Astronomers speculate that this fate may overtake a star in the following manner: as a star burns, mostly by fusion of its hydrogen and, later, helium atoms, the ashes of its fire in the form of other heavier elements such as silicon and iron accumulate at the centre. If this core of dead elements, no longer generating heat and pressure, should much exceed the mass of our own sun, the inexorable force of its own weight will be enough to cause its collapse in a matter of seconds to a body no larger than a few thousand cubic miles in volume, although still as heavy as a star. The birth of this extraordinary object, a neutron star, is a catastrophe of cosmic dimensions. Although the details of this and other similar catastrophic processes are still obscure, it is obvious that we have here, in the death throes of a large star, all the ingredients for a vast nuclear explosion. The stupendous amount of light, heat, and hard radiation produced by a supernova event equals at its peak the total output of all the other stars in the galaxy.

Explosions are seldom one hundred per cent efficient. When a star ends as a supernova, the nuclear explosive material, which includes uranium and plutonium together with large amounts of iron and other burnt-out elements, is distributed around and scattered in space just as is the dust cloud from a hydrogen bomb test. Perhaps the strangest fact of all about our planet is that it consists largely of lumps of fall-out from a star-sized hydrogen bomb. Even today, aeons later, there is still enough of the unstable explosive material remaining in the Earth's crust to enable the reconstitution on a minute scale of the original event.

Binary, or double, star systems are quite common in our galaxy, and it may be that at one time our sun, that quiet and well-behaved body, had a large companion which rapidly consumed its store of hydrogen and ended as a supernova. Or it may be that the debris of a nearby supernova explosion mingled with the swirl of interstellar dust and gases from which the sun and its planets were condensing. In either case, our solar system must have been formed in close conjunction with a supernova event. There is no other credible explanation of the great quantity of exploding atoms still present on the Earth. The most primitive and old-fashioned Geiger counter will indicate that we stand on fall-out from a vast nuclear explosion. Within our bodies, no less than three million atoms rendered unstable in that event still erupt every minute, releasing a tiny fraction of the energy stored from that fierce fire of long ago.

The Earth's present stock of uranium contains only 0.72 per cent of the dangerous isotope U235. From this figure it is easy to calculate that about four aeons ago the uranium in the Earth's crust would have been nearly 15 per cent U235. Believe it or not, nuclear reactors have existed since long before man, and a fossil natural nuclear reactor was recently discovered in Gabon, in Africa. It was in action two aeons ago when U235 was only a few per cent. We can therefore be fairly certain that the geochemical concentration of uranium four aeons ago could have led to spectacular displays of natural nuclear reactions. In the current fashionable denigration of technology, it is easy to forget that nuclear fission is a natural process. If something as intricate as life can assemble by accident, we need not marvel at the fission reactor, a relatively simple contraption, doing likewise.

Thus life probably began under conditions of radioactivity far more intense than those which trouble the minds of certain present-day environmentalists. Moreover, there was neither free oxygen nor ozone in the air, so that the surface of the

Earth would have been exposed to the fierce unfiltered ultra-violet radiation of the sun. The hazards of nuclear and of ultra-violet radiation are much in mind these days and some fear that they may destroy all life on Earth. Yet the very womb of life was flooded by the light of these fierce energies.

There is no paradox here. The present dangers are real but tend to be exaggerated. These rays are part of the natural environment and always have been. When life was first developing, the destructive bond-severing power of nuclear radiation may even have been beneficial, since it must have hastened the essential process of trial and error by dismantling the mistakes and regenerating the basic chemical spare parts. Above all, it would have hastened the production of random new combinations until the optimum form emerged.

As Urey has taught us, the Earth's primaeval atmosphere would have been blown away during the early stages when the sun was settling down. Our planet may have been for a while as bare as the moon is now. Later, the pressure of the Earth's own mass and the pent-up energy of its highly radioactive contents heated up the interior until gases and water vapour escaped to form the air and the oceans. We do not know how long it took to produce this secondary atmosphere, nor do we have evidence of its original composition, but we surmise that at the time when life began the gases from the interior were richer in hydrogen than those which now vent from volcanoes. The organic compounds, component parts of life, require that some hydrogen is available in the environment both for their formation and for their survival.

When we consider the elements from which the compounds of life are made, we usually think first of carbon, nitrogen, oxygen, and phosphorus, and then of a miscellany of trace elements, including iron, zinc, and calcium. Hydrogen, that ubiquitous material from which most of the universe is made and which occurs in all living matter, is more often taken for granted. Yet its importance and versatility are paramount. It is an essential part of any compound formed by the other key elements of life. As the fuel which powers the sun, it is the primary source of that generous flux of free solar energy which enables life's processes to start and keep going. Water, another life-essential material which is so common that we tend to forget it, is two-thirds hydrogen in atomic proportion. The abundance of free hydrogen on a planet sets the reduction-oxidation, or redox, potential, which is a measure of the tendency of an environment to oxidize or reduce. (In an oxidizing environment an element takes up oxygen, thus iron rusts. In a reducing, hydrogen-rich, environment an oxide compound tends to shed its oxygen load, thus rust turns back to iron.) The abundance of positively charged hydrogen atoms also sets the balance between acid and alkaline, or as a chemist would call it, the pH. The redox potential and the level of pH are two key environmental factors which determine whether a planet is fair or foul for life.

The American Viking spacecraft which landed on Mars and the Russian spaceship Venera which landed on Venus have both reported that no life can be seen. Venus has by now lost almost all her hydrogen and is in consequence hopelessly barren. Mars still has water and therefore chemically bound hydrogen, but its surface is so oxidized as to be bereft of the organic molecules from which life might be built. Both planets are not only dead but now never could bear life.

Although we have very little direct evidence about the chemistry of the Earth when life began, we know that it more nearly resembled the present chemical state of the giant outer planets, Jupiter and Saturn, than that of Mars or Venus. It is probable that aeons ago Mars, Venus, and the Earth had similar compositions rich in the molecules of methane, hydrogen, ammonia, and water from which life can form; but just as iron rusts and rubber perishes, time, the great oxidizer, ensures that even a planet will wither and become barren as that life-essential element hydrogen escapes to space.

The Earth must therefore have had a hydrogen-bearing, reducing atmosphere at the time when life began. This atmosphere need not have contained a high level of free hydrogen so long as outgassing provided a plentiful supply. The presence of hydrogen in molecular carriers such as methane and ammonia would have sufficed. Similar atmospheres are still to be found on the moons of the outer planets. These bodies are cold enough for their weak gravitational fields to retain a hydrogen-bearing atmosphere. Unlike these outer moons and their planets, the Earth, Mars, and Venus have neither the strong gravitational pull nor the low temperature required to retain hydrogen indefinitely without biological assistance. Hydrogen is the smallest and lightest of all atoms and so at any given temperature moves the fastest. Thus hydrogen gas at the outer edge of our atmosphere will be split into hydrogen atoms by the sun's rays. These atoms have motions rapid enough to escape the pull of the Earth's gravity and so are lost to space. This escape process will have set a time limit for the evolution of life on the Earth, since the supply of hydrogen provided by the effluent gases methane and ammonia would not have been sufficient to bridge the gap indefinitely. These gases had another vital role to play. Their presence in the atmosphere would have acted as a blanket keeping our planet warm at a time when the sun was probably less radiant than it is now.

The history of the Earth's climate is one of the more compelling arguments in favour of Gaia's existence. We know from the record of the sedimentary rocks that for the past three and a half aeons the climate has never been, even for a short period, wholly unfavourable for life. Because of the unbroken record of life, we also know that the oceans can never have either frozen or boiled. Indeed, subtle evidence from the ratio of the different forms of oxygen atoms laid down in the rocks over the course of time strongly suggests that the climate has always been much as it is now, except during glacial periods or near the beginning of life when it was somewhat warmer. The glacial cold spells—Ice Ages, as they are called, often with exaggeration—affected only those parts of the Earth outside latitudes 45° North and 45° South. We are inclined to overlook the fact that 70 per cent of the Earth's surface lies between these latitudes. The so-called Ice Ages only affected the plant and animal life which had colonized the remaining 30 per cent, which is often partially frozen even between glacial periods, as it is now.

We may at first think that there is nothing particularly odd about this picture of a stable climate over the past three and a half aeons. The Earth had no doubt long since settled down in orbit around that great and constant radiator, the sun, so why should we expect anything different? Yet it is odd, and for this reason. Our sun, being a typical star, has evolved according to a standard and well established pattern. A consequence of this is that during the three and a half aeons of life's existence on the Earth, the sun's output of energy will have increased by at least 30 per cent. Thirty per cent less heat from the sun would imply a mean temperature for the Earth well below the freezing point of water. If the Earth's climate were determined solely by the output from the sun, our planet would have been in a frozen state during the first one and a half aeons of life's existence. We know from the record of the rocks and from the persistence of life itself that no such adverse conditions existed.

If the Earth were simply a solid inanimate object, its surface temperature would follow the variations in solar output. No amount of insulating clothing will indefinitely protect a stone statue from winter cold or summer heat. Yet somehow, through three and half aeons, the surface temperature has remained constant and favourable for life, much as our body temperatures remain constant whether it is summer or winter and whether we find ourselves in a polar or tropical environment. It might be thought that the fierce radioactivity of early days would be enough to keep the planet warm. In fact simple calculations based on the very predictable nature of radioactive decay indicate that although these energies keep the interior incandescent, they have little effect on surface temperatures.

Planetary scientists have offered several explanations of our constant climate. Carl Sagan and his collaborator Dr Mullen, for example, have recently suggested that in earlier times, when the sun was dimmer, the presence of gases such as ammonia in the air helped to conserve such heat as the Earth received. Certain gases, like carbon dioxide and ammonia, absorb infra-red heat radiation from the Earth's surface and delay its escape to outer space. They are the gaseous equivalent of warm clothing. They have the additional advantage over clothing of being transparent to the incoming visible and near infra-red radiation of the sun which conveys to the Earth nearly all of the heat it receives. For this reason, although

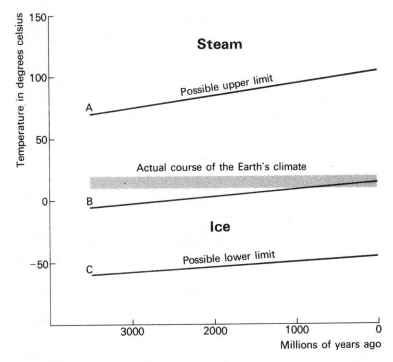

Fig. 1. The course of the Earth's average temperature since the beginning of life 3.5 aeons ago is all within the narrow bounds of the horizontal lines between 10° and 20°C. If our planetary temperature depended only on the abiological constraints set by the sun's output and the heat balance of the Earth's atmosphere and surface, then the conditions of either the upper or lower extremes, marked by the lines A and C, could have been reached. Had this happened, or even if a middle course were followed, line B, which passively goes with the sun's heat output, all life would have been eliminated.

perhaps not quite correctly, they are often called 'greenhouse' gases.

Other scientists, especially Professor Meadows and Mr Henderson Sellers of Leicester University, have suggested that in earlier times the Earth's surface was darker in colour and therefore absorbed more of the sun's heat than it does now. The proportion of sunlight reflected to space is called the albedo, or whiteness, of a planet. If its surface is completely white, it will reflect all sunlight to space and will be very cold. If it is completely black, all sunlight will be absorbed and it will be warm. A change in the albedo could obviously compensate for the lesser heat of a dimmer sun. At the present time the Earth's surface is appropriately of an intermediate colour and half covered with clouds. It reflects about 45 per cent of the incoming sunlight.

It was warm and comfortable for embryo life, in spite of the weaker flux of heat from the sun. The only explanations offered to account for this 'unseasonal winter warmth' are protection by the 'greenhouse' gases carbon dioxide and ammonia, or a lower albedo due to a different distribution of the Earth's land masses at that time. Both are possible explanations up to a point. It is where they break down that we catch our first glimpse of Gaia, or at least of the need to postulate her existence.

Once life began, it probably established itself in the sea, in the shallow waters, the estuaries, river banks, and wet lands. From these earliest habitable regions it spread to encircle the globe. When the first biosphere evolved, the chemical environment of the Earth inevitably began to change. Like the nutrients in a hen's egg, the abundant organic chemicals from which life first evolved would have supplied the infant creature with the food needed for its early growth. Unlike the chick, however, for life there was only a limited supply of food beyond the 'egg'. As soon as vital key compounds grew scarce, the infant would have been faced with the choice of starving or learning to synthesize its own building blocks from the more basic raw materials of the environment, using sunlight as the driving force.

The need to make choices of this kind must have occurred many times and hastened the diversification, independence, and sturdiness of the expanding biosphere. It may also have been during this time that the idea of predator and prey and of food chains first evolved. The natural death and decay of organisms would have released key materials to the community at large, but some species may have found it more convenient to gather their essential components by feeding on the living. The science of ecology has developed to the stage where it can now be demonstrated, with the aid of numerical models and computers, that a diverse chain of predators and prey is a more stable and stronger ecosystem than a single self-contained species, or a small group of very limited mix. If these findings are true, it seems likely that the biosphere diversified rapidly as it evolved.

One important consequence of this ceaseless activity of life would be the cycling through the biosphere of the atmospheric gases ammonia, carbon dioxide, and methane. When other sources of supply grew short these gases would supply those life-essential elements carbon, nitrogen, and hydrogen. As a result, there would be a decline in the atmospheric abundance of these gases; carbon and nitrogen would be fixed and deposited on the sea floor as organic detritus, or possibly as calcium and magnesium carbonate included in these early creatures. Some of the hydrogen released by the breakdown of ammonia would transfer to other elements, principally to oxygen to form water, and some would form hydrogen gas itself and escape into space. Nitrogen from ammonia would be left in the atmosphere in its present almost inert form as molecular nitrogen gas.

These processes may have been slow on our time-scale, but before many tenths of an aeon had elapsed the composition of the atmosphere would have changed considerably as the ammonia and carbon dioxide content was gradually depleted. If the planet was kept warm, in spite of the weaker sun, by the blanket effect of these gases, a fall in surface temperature would surely have followed this depletion. Sagan and Mullen have proposed that the climatic *status quo* may have been maintained by the biosphere learning to synthesize and replace the ammonia which it removed as food. If they are right, then here would be the first need of Gaia. Climates are inherently unstable. We are now fairly certain, thanks to the Yugoslav meteorologist Mihalanovich, that the recent periods of glaciation were a consequence of quite small changes in the Earth's orbit around the sun. A mere 2 per cent decrease in the heat received by one hemisphere is enough to establish an Ice Age. We now begin to see the awesome consequences to the infant biosphere of feeding on the atmospheric blanket, for at that critical time the sun's output was not just 2 per cent but 30 per cent less than it is now. Let us consider what might have happened if there had been even a small perturbation, such as the 2 per cent extra cooling which now precipitates a glaciation.

A fall in temperature would reduce the thickness of the ammonia blanket, because the oceans would absorb more of this gas as their surfaces cooled, and at the same time the ammonia harvest of the biosphere would become less abundant. With less ammonia in the air, heat would escape to space and a vicious circle would be established, a state of positive feedback which would make further rapid cooling inevitable. As the temperature continued to drop, there would be less ammonia in the air to prevent its further fall, and then, to cap it all, as freezing temperatures were approached, increasing ice and snow cover would rapidly raise the Earth's albedo, and thus its reflection of sunlight to space. With a 30 per cent weaker sun, a runaway world-wide fall in temperature to well below freezing would be inevitable. The Earth would become a white frozen sphere, stable and dead.

If on the other hand the biosphere had over-compensated for feeding on the atmospheric blanket and had performed its task of ammonia synthesis too well, even with a weaker sun runaway heating could take place, with the same vicious circle in reverse. The warmer it grew, the more ammonia in the air and the less heat lost to space. As the temperature increased, water vapour and additional heat-retaining gases would build up in the atmosphere. Planetary conditions would finally resemble those of Venus as she is today, although less hot. The temperature might approach 100° C, well above the toleration limit of life, and again we would have a stable but dead planet.

It may be that the natural negative feedback process of cloud formation or some other as yet unknown phenomenon would have preserved a regime at least tolerable for life, but if these fail-safe devices were not available, the biosphere would have to learn by trial and error the art of controlling its environment, at first within broad bounds and later, as control was refined, by maintaining it near the optimum state for life.

There would be more to it than simply making enough ammonia to replace the quantity consumed. It would be necessary to develop the means of sensing the temperature and the amount of ammonia in the air, so that production could be kept at the right level. This development of an active control system, however rudimentary, by the biosphere may have been the first indication that Gaia had emerged from the complex of parts.

If we are prepared to consider the biosphere as being able, like most living things, to adapt the environment to its needs, there are many ways in which these early critical climatic problems might have been solved. Most creatures can adapt their colouring for the purpose of camouflage, warning, or display. As ammonia was depleted or as the continents drifted to unfavourable positions which raised the albedo, it may have been possible for the biosphere to have kept itself and the Earth warm simply by darkening. Awramik and Golubic of Boston University have observed that on salt marshes, where the albedo is normally high, lighter-coloured carpets of micro-organisms have turned black, as the seasons changed. Could these black mats, produced by a life form with a long ancestry, be living reminders of an ancient method of conserving warmth?

Conversely, if over-heating were the cause of trouble, a marine biosphere would be able to control evaporation by producing a monomolecular layer with insulating properties to cover the surface of the waters. If evaporation from the warmer regions of the oceans were hindered by this means, it would prevent the excessive accumulation of water vapour in the atmosphere and the conditions of runaway heating by infra-red absorption.

These are examples of devices by which a biosphere might actively keep the environment comfortable. Investigations of simpler systems such as a beehive or a man suggest that temperature control would probably operate through the combined application of many different techniques rather than through any single one. The true history of these very remote periods will never be known. We can only speculate on the basis of probability and in the near-certainty that life did persist and enjoyed an equable climate.

The biosphere's first exercise in actively modifying its environment may have been concerned with climate and the cooler sun, but there are other important environmental properties which have to be kept in subtle balance if life is to persist. Some essential elements are required in bulk, others in trace quantities, and all may need rapid redeployment at times; poisonous wastes and litter must be dealt with and, if possible, put to good use; acidity must be kept in check and a neutral to alkaline overall environment maintained; the seas should stay salt, but not too salt; and so on. These are the main criteria, but there are many others involved.

As we have seen, when the first living system established itself, it was able to take advantage of the abundant supply of key components in its immediate environment, and subsequently the growing biosphere learned to synthesize these components from the basic raw materials of the air, sea, and the Earth's crust. Another essential task, as life spread and diversified, would be to ensure a reliable supply of the trace elements needed for specific mechanisms and functions. All living creatures of cellular form employ a vast array of chemical processors, or catalytic agents, called enzymes. Many of these need trace amounts of certain elements to enable them to work effectively. Thus one enzyme, carbonic anhydrase, assists in the transfer of carbon dioxide to and from the cell's environment, but this enzyme needs zinc for its formation. Other enzymes need iron, magnesium, or vanadium. Trace amounts of many other elements, including cobalt, selenium, copper, iodine, and potassium are essential for various activities in our present biosphere. No doubt similar needs arose and had to be met in the past.

At first, these trace elements would have been gathered in the usual way by drawing on the environmental bank, but in time, as life proliferated, competition for the rarer elements would have reduced the supply and checked further expansion. If, as seems likely, the shallow waters of the Earth teamed with early life, some of these key elements may have been removed from active use by the downfall of dead cells and skeletons to the muds and oozes of the sea floor. Once deposited, this detritus is usually trapped and buried in other sediments and the vital trace material is lost to the biosphere until the burial grounds are opened up by the slow, intermittent heaving of the Earth's crust. The great beds of sedimentary rocks throughout geological history bear witness to the power of this process of sequestration.

Life no doubt dealt with this problem of its own making in its own way, by the restless process of evolutionary trial and error, until a species of scavengers emerged to make a living by extracting the precious key elements from dead bodies before their burial. Other systems would have evolved intricate chemical and physical nets with which to harvest scarce materials from the sea. In time, these independent salvage operations would be merged and co-ordinated in the interests of greater productivity. The more complex co-operative network would have properties and powers greater than the sum of its parts and to this extent may be recognized as one of the faces of Gaia.

In our own society since the Industrial Revolution we have encountered major chemical problems of shortages of essential materials and of local pollution. The early biosphere must have been faced with similar problems. Perhaps the first ingenious cellular system which learned to gather zinc from its environment, at first for its own benefit and then for the common good, also unwittingly garnered the similar but poisonous element, mercury. Some slip-up of this nature probably led to one of the world's earliest pollution incidents. As usual, this particular problem was solved by natural selection, since we now have systems of micro-organisms which can convert mercury and other poisonous elements to their volatile methyl derivatives. These organisms may represent life's most ancient process for the disposal of toxic waste.

Pollution is not, as we are so often told, a product of moral turpitude. It is an inevitable consequence of life at work. The second law of thermodynamics clearly states that the low entropy and intricate, dynamic organization of a living system can only function through the excretion of low-grade products and low-grade energy to the environment. Criticism is only justified if we fail to find neat and satisfactory solutions which eliminate the problem while turning it to advantage. To grass, beetles, and even farmers, the cow's dung is not pollution but a valued gift. In a sensible world, industrial waste would not be banned but put to good use. The negative, unconstructive response of prohibition by law seems as idiotic as legislating against the emission of dung from cows.

A far more serious threat to the health of the early biosphere would have been the growing disturbances in the properties of the planetary environment as a whole. If ammonia was indeed a key primaeval gas, its consumption by the biosphere affected not only the radiation properties of the atmosphere but the balance of chemical neutrality as well. The Earth would have grown ever more acid as ammonia was removed. The conversion of methane to carbon dioxide and of sulphides to sulphates would likewise have shifted the balance towards a more acid state which life could not tolerate. We do not know how this problem was solved, but we do know that for as far back as we can measure, the Earth has been close to its present state of chemical neutrality. Mars and Venus, on the other hand, appear very acid in their composition, far too acid for life as it has evolved on our planet.

At the present time the biosphere produces up to 1,000 megatons of ammonia each year world-wide. This quantity is close to the amount required to neutralize the strong sulphuric and nitric acids produced by the natural oxidation of sulphur and nitrogen compounds: a coincidence perhaps, but possibly another link in the chain of circumstantial evidence for Gaia's existence.

A strict regulation of the salts of the ocean is as essential to life as is the need for chemical neutrality, but a much stranger and more intricate affair, as we shall see in chapter 6. Yet somehow the infant biosphere became expert in this very critical control operation as in so many others. We are bound to conclude that if Gaia does exist, the need for regulation was as urgent at the start of life as at any time since.

A well-worn shibboleth about early life is that it was shackled by a low level of available energy and that it was only when oxygen appeared in the atmosphere that evolution really took off and expanded into the full and vigorous range of life as it exists today. In fact there is direct evidence for a complex and diverse biota, one already containing all the major ecological cycles, before the appearance of skeletalized animals during the first (Cambrian) period of the Palaeozoic Era. It is true that for large mobile creatures like ourselves and certain other animals, the internal combustion of organic matter and oxygen provides a convenient source of power. But there is no biochemical reason why power need be in short supply in a reducing environment, rich in hydrogen and hydrogen-bearing molecules, so let us see how the energy game might have worked in reverse.

Some of the earliest living things have left trace fossils identified as stromatolites. These are biosedimentary structures, often laminated and shaped like cones or cauliflowers, usually composed of calcium carbonate or silica and now recognized to be products of microbial activity. Some of these are found in ancient flint-like rocks over three aeons old. Their general form suggests that they were produced by photosynthesizers, like blue-green algae of today, converting sunlight to chemical potential energy. Indeed we can be fairly sure that some early life was photosynthetic, using sunlight as the prime source of energy, for there is no other energy supply of sufficiently high potential, constancy, and quantity. The intense radioactivity of that time had the required energy potential, but in quantity it was a mere pittance compared with the sun's output.

As we have seen, the early environment of the first photosynthesizers is likely to have been a reducing one, rich in hydrogen and hydrogen-bearing molecules. The creatures of this environment could have generated just as large a chemical potential gradient for their various needs as do the plants of today. The difference would be that today oxygen is external and food and hydrogen-rich material are inside the cell, while we can speculate that aeons ago the reverse might have been true. Food for some species of primaevals may have been oxidizing substances, not necessarily free oxygen, any more than the food of present-day living cells is free hydrogen, but substances such as polyacetylenic fatty acids which release large quantities of energy when they react with hydrogen.

Fig. 2. A stromatolite colony on the shore in South Australia. This is very close in structure to the fossil remains of similar colonies from 3,000 million years ago. Photograph by P. F. Hoffman, supplied by M. R. Walter.

Strange compounds like these are still produced by certain soil micro-organisms, and are the analogues of the fats which store energy in human cells today.

This imaginary topsy-turvy biochemistry may have had no real existence. The point is that organisms with the ability to convert the energy of sunlight into stored chemical power would have had ample capacity and free energy, even in a reducing atmosphere, with which to carry out most biochemical processes.

The geological record shows that vast quantities of crustal rock containing the ferrous or more reduced form of iron were oxidized during the early stages of life. This could be evidence that the original biosphere produced hydrogen and maintained a sufficient abundance of this gas and its compounds, such as ammonia, in the atmosphere to lead to the escape of hydrogen to space. Ycas has rightly commented, in a letter to *Nature*, that biological intervention may be needed to explain the large escape of hydrogen from the Earth.

Eventually, perhaps two aeons ago, all the reducing materials of the crust were oxidized more rapidly than they were exposed geologically, and the continued activity of aerobic photosynthesizers led to the accumulation of oxygen in the air. This was probably the most critical period of all in the history of life on the Earth. Oxygen gas in the air of an anaerobic world must have been the worst atmospheric pollution incident that this planet has ever known. We need only imagine the effect on our contemporary biosphere of a marine alga which successfully colonized the sea and then used sunlight to generate chlorine from the abundant chloride ion of sea water. The devastating effect of a chlorine-bearing atmosphere on contemporary life could hardly be worse than the impact of oxygen on anaerobic life some two aeons ago.

This momentous era also marked the end of the ammonia blanket as a device to keep the planet warm. Free oxygen and ammonia react in the atmosphere and set a limit to the maximum possible abundance of ammonia gas. At the present time it is less than one part in one hundred million, far too little to exert any useful influence on infra-red absorption although, as we have seen, even this quantity of ammonia is still useful as a neutralizer of acidity, which is otherwise an inevitable by-product of oxidation, and plays its part in preventing the environment from drifting to an acid state inconsistent with life.

When oxygen leaked into the air two aeons ago, the biosphere was like the crew of a stricken submarine, needing all hands to rebuild the systems damaged or destroyed and at the same time threatened by an increasing concentration of poisonous gases in the air. Ingenuity triumphed and the danger was overcome, not in the human way by restoring the old order, but in the flexible Gaian way by adapting to change and converting a murderous intruder into a powerful friend.

The first appearance of oxygen in the air heralded an almost fatal catastrophe for early life. To have avoided by blind chance death from freezing or boiling, from starvation, acidity, or grave metabolic disturbance, and finally from poisoning, seems too much to ask; but if the early biosphere was already evolving into more than just a catalogue of species and was assuming the capacity for planetary control, our survival through those hazardous times is less difficult to comprehend.

CHAPTER 3

The recognition of Gaia

Picture a clean-swept sunlit beach with the tide receding; a smooth flat plain of golden glistening sand where every random grain has found due place and nothing more can happen.

Beaches are, of course, in reality seldom absolutely flat, smooth, and undisturbed, or at least not for long. The golden reaches of sand are continually resculptured by fresh winds and tides. Yet events may still be circumscribed. We may still be in a world where change is no more than the shifting profile of the wind-swept dunes; or no more than the ebb and flow of the tide, creating and erasing its own ripples in the sand.

Now let us suppose that our otherwise immaculate beach contains one small blot on the horizon: an isolated heap of sand which at close range we recognize instantly to be the work of a living creature. There is no shadow of a doubt, it is a sand-castle. Its structure of piled truncated cones reveals the bucket technique of building. The moat and drawbridge, with its etched facsimile of a portcullis already fading as the drying wind returns the grains to their equilibrium state, are also typical. We are programmed, so to speak, for instant recognition of a sand-castle as a human artefact, but if more proof were needed that this heap of sand is no natural phenomenon, we should point out that it does not fit with the conditions around it. The rest of the beach has been washed and brushed into a smooth carpet; the sand-castle has still to crumble; and even a child's fortress in the sand is too intricate in the design and relationship of its parts, too clearly purpose-built, to be the chance structure of natural forces.

Even in this simple world of sand and sand-castles there are clearly four states: the inert state of featureless neutrality and complete equilibrium (which can never be found in reality on Earth so long as the sun shines and gives energy to keep the air and sea in motion, and thus shift the grains of sand); the structured but still lifeless 'steady state', as it is called, of a beach of rippled sand and wind-piled dunes; the beach when exhibiting a product of life in the sand-castle; and finally the state when life itself is present on the scene in the form of the builder of the castle.

The third order of complexity represented by the sand-castle, between the abiological, or non-living, steady state and that when life is present, is important in our quest for Gaia. Though lifeless themselves, the constructions made by a living creature contain a wealth of information about the needs and intentions of their builder. The clues to Gaia's existence are as transient as our sand-castle. If her partners in life were not there, continually repairing and recreating, as children build fresh castles on the beach, all Gaia's traces would soon vanish.

How, then, do we identify and distinguish between the works of Gaia and the chance structures of natural forces? And how do we recognize the presence of Gaia herself? Fortunately we are not, like those demented hunters of the Snark, entirely without a chart or means of recognition; we have some clues. At the end of the last century Boltzman made an elegant redefinition of entropy as a measure of the probability of a molecular distribution. It may seem at first obscure, but it leads directly to what we seek. It implies that wherever we find a highly improbable molecular assembly it is probably life or one of its products, and if we find such a distribution to be global in extent then perhaps we are seeing something of Gaia, the largest living creature on Earth.

But what, you may ask, is an improbable distribution of molecules? There are many possible answers, such as the rather unhelpful ones: an ordered distribution of improbable molecules (like you, the reader), or an improbable distribution of common molecules (as, for example, the air). A more general answer, and one useful in our quest, is a distribution which is sufficiently different from the background state to be recognizable as an entity. Another general definition of an improbable molecular distribution is one which would require the expenditure of energy for its assembly from the background of molecules at equilibrium. (Just as our sand-castle is recognizably different from its uniform background, and the extent to which it is different or improbable is a measure of the entropy reduction or purposeful life-activity that it represents.)

We now begin to see that the recognition of Gaia depends upon our finding on a global scale improbabilities in the distribution of molecules so unusual as to be different and distinguishable, beyond reasonable doubt, from both the steady state and the conceptual equilibrium state.

It will help us in our quest to start with a clear idea of what the Earth would be like, both in the equilibrium state and in the lifeless steady state. We also need to know what is meant by chemical equilibrium.

The state of disequilibrium is one from which, in principle at least, it should be possible to extract some energy, as when a grain of sand falls from a high spot to a low one. At equilibrium, all is level and no more energy is available. In our small world of sand grains the fundamental particles were effectively all of the same, or of very similar, material. In the real world there are one hundred or more chemical elements with the capacity to join together in many different ways. A few of them—carbon, hydrogen, oxygen, nitrogen, phosphorus, and sulphur—are capable of interacting and interlinking to an almost infinite extent. However, we know more or less the proportions of all the elements in the air, the sea, and the surface rocks. We also know the amount of energy released when each of these elements combines with another and when their compounds combine in their turn. So if we assume that there is a constant random source of disturbance, like a fitful wind in our sand world, we can calculate what will be the distribution of chemical compounds when the state of lowest energy is reached, in other words that state from which no further energy can be gained by chemical reactions. When we do this calculation, with of course the aid of a computer, we find that our chemical equilibrium world is approximately as shown in Table 1 (p. 36).

The distinguished Swedish chemist Sillen was the first to calculate what would be the result of bringing the substances of the Earth to thermodynamic equilibrium. Many others have done so since and have substantially confirmed his work. It is

Table 1. A comparison of the composition of the oceans and the air of the present world and of a hypothetical chemical equilibrium world

	Principal components per cent		
	Substance	Present world	Equilibrium world
AIR	Carbon dioxide	0.03	99
	Nitrogen	78	0
	Oxygen	21	0
	Argon	1	1
OCEAN	Water	96	63
	Salt	3.5	35
	Sodium nitrate	traces	1.7

one of those exercises where the imagination can be set free through the assistance of a computer as a faithful and willing slave to perform the many tedious calculations.

On the scale of the Earth itself, we must swallow some formidable academic unrealities to reach the equilibrium state. We have to imagine that somehow the world has been totally confined within an insulated vessel, like a cosmic Dewar flask, kept at 15°C. The whole planet is then uniformly mixed until all possible chemical reactions have gone to completion, and the energy they released has been removed so as to keep the temperature constant. We might finish with a world covered with a layer of ocean, devoid of waves or ripples, above which there would be an atmosphere rich in carbon dioxide and devoid of oxygen and nitrogen. The sea would be very salty and the sea bed would consist of silica, silicates, and clay minerals.

The exact chemical composition and form of our chemical equilibrium world is less important than the fact that in such a world there is no source of energy whatever: no rain, no waves or tides, and no possibility of a chemical reaction which would yield energy. It is very important for us to understand that such a world—warm, damp, with all of the ingredients of life at hand—could never bear life. Life requires a constant energy flux from the sun to sustain it.

This abstract equilibrium world differs from a possibly real but lifeless Earth in the following significant ways: the real Earth would be spinning and in orbit around the sun and subject therefore to a powerful flux of radiant energy, which would include some radiation capable of splitting molecules at the atmosphere's outer reaches. It would also have a hot interior, maintained by the disintegration of radioactive elements left over from whatever cataclysmic nuclear explosion produced the debris from which the Earth was formed. There would be clouds and rain, and possibly some land. Assuming the present solar output, polar ice caps would be unlikely, for this steady-state lifeless world would be richer in carbon dioxide and consequently lose heat less readily than does the real world we live in today.

In a real but lifeless world a little oxygen might appear, as water decomposed at the outer reaches of the atmosphere and the light hydrogen atoms escaped to space. Just how much oxygen is very uncertain and a matter of debate. It would depend on the rate at which reducing materials entered from below the crust and also on how much hydrogen was returned from space. We can be sure, though, that if oxygen were present it would be no more than a trace as is found on Mars now. In this world, power would be available since windmills and water-wheels would function but chemical energy would be very hard to find. Nothing remotely like a fire could be lit. Even if traces of oxygen did accumulate in the atmosphere, there would be no fuel to burn in it. Even if fuel were available, at least 12 per cent oxygen in the air is needed to start a fire, and this is far, far more than the trace amount of a lifeless world.

Although the lifeless steady-state world differs from the imaginary equilibrium world, the difference between them is very much less than that between either of them and our living world of today. The wide differences in the chemical composition of the air, sea, and land are the subjects of later chapters. For the present, the point of interest is that everywhere on the Earth today chemical power is available and in most places a fire can be lit. Indeed, it would require only an increase of about 4 per cent in the atmospheric level of oxygen to bring the world into danger of conflagration. At 25 per cent oxygen level even damp vegetation will continue to burn once combustion has started, so that a forest fire started by a lightning flash would burn fiercely till all combustible material was consumed. Those science fiction stories of other worlds with bracing atmospheres due to the richer oxygen content are fiction indeed. A landing of the heroes' spaceship would have destroyed the planet.

My interest in fires and the availability of chemical free energy is not due to some quirk or pyromanic tendency; it is because recognizability in chemical terms can be measured by the intensity of free energy (for example, the power available from lighting a fire). By this measure alone our world, even the non-living part of it, is recognizably very different from the equilibrium and steady-state worlds. Sand-castles would vanish from the Earth in a day if there were no children to build them. If life were extinguished, the available free energy for lighting fires would vanish just as soon, comparatively, as oxygen vanished from the air. This would happen over a period of a million years or so, which are as nothing in the life of a planet.

The keynote, then, of this argument is that just as sand-castles are almost certainly not accidental consequences of natural but non-living processes like wind or waves, neither are the chemical changes in the composition of the Earth's surface and atmosphere which make the lighting of fires possible. All right, you may say, you are establishing a convincing case for the idea that many of the non-living features of our world, like the ability to light a fire, are a direct result of the presence of life, but how does this help us to recognize the existence of Gaia? My answer is that where these profound disequilibria are global in extent, like the presence of oxygen and methane in the air or wood on the ground, then we have caught a glimpse of something global in size which is able to sustain and keep constant a highly improbable distribution of molecules.

The lifeless worlds which I have modelled for comparison

with our present living world are not very sharply defined and geologists might query the distribution of elements and compounds. Certainly there is room for debate on how much nitrogen a non-living world would possess. It will be particularly interesting to learn more about Mars and its nitrogen content, and whether this gas is chemically bound on the surface as nitrate or some other nitrogen compound or whether it has, as Professor Michael McElroy of Harvard University has suggested, escaped to space. Mars may well be a prototype non-living steady-state world.

Because of these uncertainties, let us consider two other ways of constructing a steady-state lifeless world and then see how they compare with the model world we have already discussed. Let us assume that Mars and Venus are indeed lifeless and interpolate between them, in place of our present Earth, a hypothetical lifeless planet. Its chemical and physical features, in relation to its neighbours, can perhaps best be imagined in terms of a fictional country sited half-way between Finland and Libya. The atmospheric compositions of Mars, our present Earth and Venus, and of our hypothetical abiological Earth are listed in Table 2.

Another way is to assume that one of those predictions of

Table 2

Gas	Planet			
	Venus	Earth without life	Mars	Earth as it is
Carbon dioxide	98%	98%	95%	0.03%
Nitrogen	1.9%	1.9	2.7%	79%
Oxygen	trace	trace	0.13%	21%
Argon	0.1%	0.1%	2%	1%
Surface temperatures °C	477	290±50	−53	13
Total pressure bars	90	60	0064	1.0

imminent doom for our planet came true and that all life on Earth ceased, down to the last spore of some deep-buried anaerobic bacteria. So far, no doom scenario yet imagined has the slightest chance of achieving such a degree of destruction; but let us assume that it could. In order to carry out our experiment properly and trace the changing chemical scene during the Earth's transition from a healthy life-bearing world to a dead planet, we need to find a process which will remove life from the scene without altering the physical environment. Contrary to the forebodings of many environmentalists, finding a suitable killer turns out to be an almost insoluble problem. There is the alleged threat posed by aerosols to the ozone layer, which if depleted would allow a flood of lethal ultra-violet radiation from the sun to 'destroy all life on Earth'. The complete or partial removal of the ozone layer could have unpleasant consequences for life as we know it. Many species, including man, would be discomforted and some would be destroyed. Green plants, the primary producers of food and oxygen, might suffer but, as has been recently shown, some species of blue-green algae, the primary power transformers of ancient times and modern shores, are highly resistant to short-wave ultra-violet radiation. Life on this planet is a very tough, robust, and adaptable entity and we are but a small part of it. The most essential part is probably that which dwells on the floors of the continental shelves and in the soil below the surface. Large plants and animals are relatively unimportant. They are comparable rather to those elegant salesmen and glamorous models used to display a firm's products, desirable perhaps, but not essential. The tough and reliable workers composing the microbial life of the soil and sea-beds are the ones who keep things moving, and they are protected against any conceivable level of ultra-violet light by the sheer opacity of their environment.

Nuclear radiation has lethal possibilities. If a nearby star becomes a supernova and explodes, would not the flood of cosmic rays sterilize the Earth? Or what if all the nuclear weapons stock-piled on earth were exploded almost simultaneously in a global war? Again, we and the larger animals and plants might be seriously affected, but it is doubtful whether

unicellular life would for the most part even notice such an
event. There have been many investigations of the ecology of
the Bikini Atoll to see if the high level of radioactivity,
resulting from the bomb tests there, had adversely affected the
life of that coral island. The findings show that, in spite of the
continuing radioactivity in the sea and on land, this has had
little effect on the normal ecology of the area, except in places
where the explosions had blown away the top soil and left bare
rock behind.

Towards the end of 1975 the United States National
Academy of Sciences issued a report prepared by an eight-man
committee of their own distinguished members, assisted by
forty-eight other scientists chosen from those expert in the
effects of nuclear explosions and all things subsequent to them.
The report suggested that if half of all of the nuclear weapons
in the world's arsenals, about 10,000 megatons, were used in a
nuclear war the effects on most of the human and man-made
ecosystems of the world would be small at first and would
become negligible within thirty years. Both aggressor and
victim nations would of course suffer catastrophic local devastation, but areas remote from the battle and, especially important in the biosphere, marine and coastal ecosystems would be
minimally disturbed.

To date, there seems to be only one serious scientific criticism of the report, namely, of the claim that the major global
effect would be the partial destruction of the ozone layer by
oxides of nitrogen generated in the heat of the nuclear explosions. We now suspect that this claim is false and that
stratospheric ozone is not much disturbed by oxides of nitrogen. There was, of course, at the time of the report a strange
and disproportionate concern in America about stratospheric
ozone. It might in the end prove to be prescient, but then as
now it was a speculation based on very tenuous evidence. In
the nineteen-seventies it still seems that a nuclear war of
major proportions, although no less horrific for the participants and their allies, would not be the global devastation so
often portrayed. Certainly it would not much disturb Gaia.

The report itself was criticized then as now on political and
moral grounds. It was judged irresponsible as it might even

encourage the bomb-happy among the military planners to let fly.

It seems that to delete life from our planet without changing it physically is well-nigh impossible. We are left for our experiment with only science-fictional possibilities, so let us construct a doom scenario in which all life on earth down to that last deep-buried spore is indeed annihilated.

Dr Intensli Eeger is a dedicated scientist, employed by an efficient and successful agricultural research organization. He is much distressed by those appalling pictures of starving children shown in the Oxfam appeals. He is determined to devote his scientific skills and talents to the task of increasing the world's food production, especially in those underdeveloped regions which gave rise to the Oxfam pictures. His work plan is based on the idea that food production in these countries is hindered among other things by a lack of fertilizers, and he knows that the industrial nations would find it difficult to produce and deliver simple fertilizers such as nitrates and phosphates in sufficient quantities to be of any use. He also knows that the use of chemical fertilizers alone has drawbacks. He plans instead to develop by genetic manipulation a greatly improved strain of nitrogen-fixing bacteria. By this means, nitrogen in the air could be transferred directly to the soil without the need for a complex chemical industry, and also without disturbing the natural chemical balance of the soil.

Dr Eeger had spent many years of patient trial with many promising strains which did wonders on the laboratory field plots but which failed when transferred to the tropical test grounds. He persisted until one day he heard by chance from a visiting agriculturalist that a strain of maize had been developed in Spain which flourished in soil poor in phosphate. Dr Eeger had a hunch. He guessed that maize would be unlikely to flourish in such a soil without assistance. Was it possible that it had acquired a co-operative bacterium which, like that which lives on the roots of clover and can fix nitrogen from the air, had somehow contrived to gather what phosphate there was in the soil for the benefit of the maize?

Dr Eeger spent his next holiday in Spain, close to the agricultural centre where the work on maize was being done, having previously arranged to visit his Spanish colleagues and discuss the problem. They met and talked, and exchanged samples. On returning to his laboratory, Dr Eeger cultivated the maize and from it extracted a motile micro-organism with a capacity to gather phosphate from soil particles far more efficiently than any other organism he had ever known. It was not difficult for a man of his skills to contrive the adaptation of this new bacterium so that it could live comfortably with many other food crops and particularly with rice, the most important food source in the tropical regions. The first trials of cereals treated with *Phosphomonas eegarii* at the English test site were astonishingly successful. Yields of all the crops they tried were substantially increased. Moreover no harmful or adverse effects were found in any of the tests.

The day came for the tropical trial at the field station in Northern Queensland. A culture of *P. eegarii* was without ceremony sprayed in diluted form upon a small patch of experimental rice paddy. But here the bacterium forsook its contrived marriage with the cereal plants and formed a more exciting but adulterous union with a tough and self-sufficient blue-green alga growing on the water surface of the paddy field. They grew happily together, doubling in numbers every twenty minutes in the warm tropical environment, the air and soil providing all they needed. Small predatory organisms would normally have ensured a check on such a development, but this combination was not to be stopped. Its capacity to gather phosphorus rendered the environment barren for everything else.

Within hours, the rice paddy and those around it took on the appearance of a ripe duck pond covered with lurid iridescent green scum. It was realized that something had gone badly wrong and the scientists soon uncovered the association of *P. eegarii* and the alga. Foreseeing the dangers with rare promptness, they arranged that the entire paddy area and the water channels leading from it be treated with a biocide and the growth destroyed.

That night, Dr Eeger and his Australian colleagues went late to bed, tired and worried. The dawn fulfilled their worst fears. The new bloom, like some living verdigris, covered the surface of a small stream a mile away from the paddies and only a few miles from the sea. Again, every agent of destruction was applied wherever the new organism might have travelled. The director of the Queensland station tried desperately but in vain to persuade the government to evacuate the area at once and use a hydrogen bomb to sterilize it before the spread was beyond all possibility of control.

In two days the algal bloom had started to spread into the coastal waters, and by then it was too late. Within a week the green stain was clearly visible to airline passengers flying six miles above the Gulf of Carpentaria. Within six months more than half of the ocean and most of the land surfaces were covered with a thick green slime which fed voraciously on the dead trees and animal life decaying beneath it.

By this time Gaia was mortally stricken. Just as we all too frequently die through the uncontrolled growth and spread of an errant version of our own cells, so the cancerous algal-bacterial association had displaced all the intricate variety of cells and species which make up the healthy living planet. The near-infinity of creatures performing essential co-operative tasks was displaced by a greedy, uniform green scum, knowing nothing but an insatiable urge to feed and grow.

Viewed from space, the Earth had changed to a blotchy green and faded blue. With Gaia moribund, the cybernetic control of the Earth's surface composition and atmosphere at an optimum value for life had broken down. Biological production of ammonia had long ceased. Decaying matter, including vast quantities of the alga itself, produced sulphur compounds that oxidized to sulphuric acid in the atmosphere. So the rain fell ever more acid upon the land and steadily denied that habitat to the usurper. The lack of other essential elements began to exert its effect and gradually the algal bloom faded until it survived only in a few marginal habitats, where nutrients were for a while still available.

Now let us see how this stricken Earth would move slowly but inexorably towards a barren steady state, although the time scale might be of the order of a million years or more. Thunderstorms and radiation from the sun and space would continue to bombard our defenceless world and would now sever the more stable chemical bonds, enabling them to recombine in forms closer to equilibrium. At first, the most important of these reactions would be that between oxygen and the dead organic matter. Half of it might be oxidized, while the rest would be covered with mud and sand and buried. This process would remove only a small percentage of the oxygen. More slowly and surely it would also combine with the reduced gases from volcanoes and with the nitrogen of the air. As the nitric and sulphuric acid rain washed the earth, some of the vast store of carbon dioxide fixed by life as limestone and chalk would be returned as gas to the atmosphere.

As explained in the previous chapter, carbon dioxide is a 'greenhouse' gas. With small quantities its effect on the temperature of the air is proportional to the amount added or, as the mathematicians would say, there is a linear effect. However, once the carbon dioxide concentration in the air approaches or exceeds 1 per cent, new non-linear effects come into play and the heating greatly increases. In the absence of a biosphere to fix carbon dioxide, its concentration in the atmosphere would probably exceed the critical figure of 1 per cent. The Earth would then heat up rapidly to a temperature near to that of boiling water. Increasing temperatures would speed up chemical reactions and accelerate their progression towards chemical equilibrium. Meantime, all traces of our algal destroyer would finally have vanished, sterilized by the boiling seas.

In our present world, the very low temperatures at about seven miles above the Earth's surface freeze out water vapour until there is only about one part in a million left. The escape of this tiny portion upwards, where it may dissociate to produce oxygen, is so slow as to be of no consequence. However, the violent weather of a world of boiling seas would probably generate thunder clouds penetrating as far as the upper atmosphere and causing there a rise in temperature and humidity. This in turn might encourage the more rapid decomposition of water, the escape of hydrogen so formed into space, and a greater production of oxygen. The release of more oxygen would ensure the ultimate removal of virtually all nitrogen from the air. The atmosphere would eventually consist of carbon dioxide and steam, with a little oxygen (probably less than 1 per cent) and the rare gas argon and its relatives, which have no chemical role to play. The Earth would become permanently cocooned in brilliant white cloud—a second Venus, although not quite as hot.

The run-down to equilibrium could follow a very different path. If, during its period of insatiable growth, the alga had greatly depleted atmospheric carbon dioxide, the Earth might have been set on a course of irreversible cooling. Just as an excess of carbon dioxide leads to overheating, so its removal from the atmosphere could lead to runaway freezing. Ice and snow would cover most of the planet, freezing to death the last of that over-ambitious life form. The chemical combination of nitrogen and oxygen would still take place, but rather more slowly. The end result would be a more or less frozen planet with a thin low-pressure atmosphere of carbon dioxide and argon, and with mere traces of oxygen and nitrogen. In other words, like Mars, although not quite as cold.

We cannot be certain which way things would go. What is certain is that with Gaia's intelligence network and intricate system of checks and balances totally destroyed, there would be no going back. Our lifeless Earth, no longer a colourful misfit, a planet that broke all the rules, would fall soberly into line, in barren steady state, between its dead brother and sister, Mars and Venus.

It is necessary for me to remind you that the foregoing is fiction. It may be scientifically plausible as a model but then only if the postulated bacterial association could exist, remain stable, and exert its aggression without check or hindrance. The genetic manipulation of micro-organisms for the benefit of mankind has been a busy activity ever since they were domesticated for such purposes as cheese- and wine-making. As everyone who is a practitioner of these arts and indeed every farmer will confirm, domestication does not favour survival under wild conditions. So strongly expressed, however, has been public concern over the dangers of genetic manipulations involving DNA itself, that it was good to have no less an authority than John Postgate confirm that this brief essay in science fiction is indeed just a flight of fancy. In real life, there must be many taboos written into the genetic coding, the universal language shared by every living cell. There must also be an intricate security system to ensure that exotic outlaw species do not evolve into rampantly criminal syndicates. Vast numbers of viable genetic combinations must have been tried out, through countless generations of micro-organisms, during the history of life.

Perhaps our continuing orderly existence over so long a period can be attributed to yet another Gaian regulatory process, which has evolved to maintain internal genetic security.

CHAPTER 4

Cybernetics

The American mathematician Norbert Wiener first gave common use to the word 'cybernetics' (from the Greek word for 'steersman', 'kubernetes'), to describe that branch of study which is concerned with self-regulating systems of communication and control in living organisms and machines. The derivation seems apt, since the primary function of many cybernetic systems is to steer an optimum course through changing conditions towards a predetermined goal.

We know from long experience that stable objects are those with broad bases and with most of their mass centred low, yet we seldom marvel at our own remarkable ability to stand upright, supported only by our jointed legs and narrow feet. To stay erect even when pushed, or when the surface beneath us moves, as on a ship or a bus; to be able to walk or run over rough ground without falling; to keep cool when it is hot or vice versa, are examples of cybernetic processes and of properties exclusive to living things and to highly automated machines.

We stand upright—after a little practice—on a ship that rolls because we possess an array of sensory nerve cells buried in our muscles, skin, and joints. The function of these sensors is to provide a constant flow of information to the brain about the movements and location in space of the various parts of our bodies, as well as the environmental forces currently acting on them. We also have a pair of balance organs associated with our ears which work like spirit-levels, each having a bubble moving in a fluid medium to record any change in the position of the head; and we have our eyes to scan the horizon and tell us how we stand in relation to it. All this flow of information is processed by the brain, usually at an unconscious level, and is immediately compared with our consciously intended stance at the time. If we have decided to stand level in spite of the ship's motion, perhaps to look at the receding harbour through binoculars, this chosen posture is the reference point used by the brain to compare with departures from it caused by the rolling of the vessel. Thus our sense organs continually inform the brain about our stance, and counter-instructions pass constantly from the brain down the motor nerves to the muscles. As we tip from the vertical, the push-and-pull of these muscles changes, so as continuously to maintain the upright position.

This process of comparing wish with actuality, of sensing error and then correcting it by the precise application of an opposing force enables us to stand erect. Walking or balancing on one leg is more difficult and takes longer to learn; riding a bicycle is even trickier, but this also can become second nature through the same active control process which keeps us upright.

It is worth emphasizing the subtle mechanisms involved in simply standing in one place. If, for instance, when the deck beneath us tilts slightly, the correcting force applied by our muscles is too great, we shall be driven too far the other way; over-anxious compensation might then swing us suddenly in the original direction of the tilt, setting up an oscillation which could topple us or at the least frustrate our wish to stay upright. Such instabilities or oscillations in cybernetic systems are all too common. There is a pathological condition known as 'intention tremor'. With this the unfortunate sufferer who tries to pick up a pencil overreaches his target, over-compensates and swings too far the other way, oscillating back and forth in the frustrating failure to achieve a simple aim. It is not enough merely to oppose a force which is pushing us away from our goal; we must, smoothly, precisely, and continuously, match the power of the opposition if we are to keep to our purpose.

What, you may be wondering, has all this to do with Gaia? Possibly a great deal. One of the most characteristic properties of all living organisms, from the smallest to the largest, is their capacity to develop, operate, and maintain systems which set a goal and then strive to achieve it through the cybernetic process of trial and error. The discovery of such a system, operating on a global scale and having as its goal the establishment and maintenance of optimum physical and chemical conditions for life, would surely provide us with convincing evidence of Gaia's existence.

Cybernetic systems employ a circular logic which may be unfamiliar and alien to those of us who have been accustomed to think in terms of the traditional linear logic of cause and effect. So let us start by considering some simple engineering systems which employ cybernetics to maintain a chosen state. Take temperature control, for example. Most homes now possess a cooking oven, an electric iron, and a heating system for the house. In each of these devices the goal is to maintain the desired and appropriate temperature. The iron must be hot enough to smooth without scorching; the oven to cook rather than burn or undercook; and the heating system is required to keep the house comfortably warm and neither too hot nor too cold. Let us examine the oven more closely. It consists of a box, designed to conserve heat without losing it too fast to the kitchen, a control panel, and heating elements which convert the electrical power supply into heat within the oven. Inside the oven there is a special kind of thermometer called a thermostat. This device does not need to show the temperature visually, like an ordinary room thermometer. It is arranged instead to operate a switch when the desired temperature is reached. This chosen temperature is set and indicated by a dial on the control panel, which is linked directly with the thermostat. An essential and perhaps surprising feature of a well-designed oven is that it must be capable of reaching temperatures far higher than will ever be needed in cooking: otherwise, the time taken to reach the desired level of heat would be far too long. If, for example, the dial is set at 300 degrees and the oven is turned on, the elements will be supplied with full power and will often glow red-hot, rapidly flooding the box interior with heat. The temperature rises fast until the thermostat recognizes that the set level of 300 degrees has been reached. The power supply is then switched off, but the inside temperature continues to rise for a short while, as heat flows from the red-hot elements. As they cool, the temperature drops, and when the thermostat senses that it has fallen below 300 degrees the power is switched on again.

There is a brief period of further cooling while the heaters warm and then the cycle recommences. The oven temperature thus swings a few degrees above and below the desired level. This small margin of error in temperature control is a characteristic feature of cybernetic systems. Like living things, they seek or approach perfection but never quite make it.

Now what is so special about this arrangement? Grandmother was surely able to cook magnificent meals without using a newfangled oven equipped with a thermostat. But was she? True, in grandmother's day the oven was heated by burning wood or coal and so arranged that if all went well just enough heat from the fire reached the oven to keep it at the right temperature. Yet on its own such an oven could never cook properly; it would either burn the cakes or leave them sad and stodgy. Its efficiency depended entirely on grandmother herself functioning as the thermostat. She learned to read the oven's signs and recognize when the desired temperature was reached; she knew that it was then time to damp down the fire. At intervals she would check that the food was cooking satisfactorily, judging by sound and smell as well as sight and feel. Today, an engineer could design an oven just as good with a robot grandmother to sit in the kitchen and watch it, sensing its temperature and remotely controlling the electricity supply.

Anyone who tries to cook by an oven lacking either human or mechanical supervision soon finds that the results are far from satisfactory. To maintain the required temperature for, say, an hour, it is essential that the input of heat compensates exactly for any heat losses from the oven. A cold draught from outside, a change in the electricity supply voltage or gas pressure, the size of the meal to be cooked, and whether or not other parts of the stove are in use, are all factors which could frustrate our desire to attain the right working temperature for the right length of time.

The attainment of any skill, whether it be in cooking, painting, writing, talking, or playing tennis, is all a matter of cybernetics. We aim at doing our best and making as few mistakes as possible; we compare our efforts with this goal and learn by experience; and we polish and refine our performance by constant endeavour until we are satisfied that we are as near to optimum achievement as we can ever reach. This process is well called learning by trial and error.

It is interesting to recall that well into the nineteen-thirties men and women were using cybernetic techniques throughout their lives without conscious recognition. Engineers and scientists were applying them to the design of intricate instruments and mechanical devices. Yet nearly all these activities were performed without a formal understanding or logical definition of what was involved. It was rather like Monsieur Jourdain, Molière's would-be gentleman, who had never realized that what he spoke was prose. The over-long delay in the understanding of cybernetics is perhaps another unhappy consequence of our inheritance of classical thought processes. In cybernetics, cause and effect no longer apply; it is impossible to tell which comes first, and indeed the question has no relevance. The Greek philosophers abhorred a circular argument as firmly as they believed that nature abhorred a vacuum. Their rejection of circular arguments, the key to understanding cybernetics, was as erroneous as their assumption that the universe was filled with the air we breathe.

Think again about our temperature-controlled oven. Is it the supply of power that keeps it at the right temperature? Is it the thermostat, or the switch that the thermostat controls? Or is it the goal we established when we turned the dial to the required cooking temperature? Even with this very primitive control system, little or no insight into its mode of action or performance can come from analysis, by separating its component parts and considering each in turn, which is the essence of thinking logically in terms of cause and effect. The key to understanding cybernetic systems is that, like life itself, they are always more than the mere assembly of constituent parts. They can only be considered and understood as operating systems. A switched off or dismantled oven reveals no more of its potential performance than does a corpse of the person it once was.

The Earth spins before an uncontrolled radiant heater, the sun, whose output is by no means constant. Yet right from the beginning of life, around three and a half aeons ago, the

Earth's mean surface temperature has never varied by more than a few degrees from its current levels. It has never been too hot or too cold for life to survive on our planet, in spite of drastic changes in the composition of the early atmosphere and variations in the sun's output of energy.

In chapter 2 I discussed the possibility that the Earth's surface temperature is actively maintained at an optimum by and for the complex entity which is Gaia, and has been so maintained for most of her existence. What parts of herself, I wonder, does she use as the thermostat? It is unlikely that a single simple control mechanism for planetary temperature would be subtle enough to serve her purpose. Moreover, three and a half aeons of experience and of research and development have no doubt given time and opportunity for the evolution of a highly sophisticated and comprehensive control system. We shall have some notion of the subtleties we need to look for and may expect to find during the disentanglement of Gaia's mechanism for temperature regulation if we consider how the temperature of our own bodies is regulated for us.

The clinical thermometer still provides the physician with evidence for or against a suspected invasion by foreign micro-organisms, and the pattern of variations in the rise and fall of the patient's temperature which it reveals gives him useful information as to the identity of the invaders. In fact, it has been so invaluable as a diagnostic aid that some diseases, such as undulant fever, are named after their characteristic temperature patterns. Yet to nearly all physicians, even today, the processes by which the body controls its temperature are as mysterious as they are to their patients. It is only in recent years that some physiologists, showing great courage and mental stamina, have given up their work in medicine to retrain as systems engineers. From this new beginning has come partial understanding of the wonderfully co-ordinated process of body temperature regulation.

Our temperature in health is not maintained at a constant value, that mythical normal level of 98.4° F (37°C); it varies according to the needs of the moment. If we are obliged to run or exercise continuously, it will rise by several degrees, well into the fever area. In the early hours or when we starve it may fall as far below 'normal'. Moreover, this relatively constant value of 98.4° F only applies to our core region, which covers the trunk and the head, wherein lie most of the important administrative systems of the body. Our skin, hands, and feet have to endure a greater range of temperature, and are designed to function even when near to freezing with no more than a shiver of complaint.

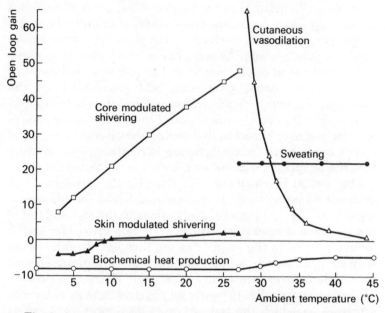

Fig. 3. An engineer's diagram illustrating the power of the five processes of human temperature regulation to function when a naked man is exposed to different environmental temperatures.

T. H. Benzinger and his colleagues extended the horizon with their discovery that body temperature is kept at a continuous optimum by a consensus decision arrived at by the brain in consultation with other parts of the body as to the most suitable temperature for the occasion. The reference is not so much to temperature scale, but to the efficiency range of the different organs of the body in relation to body temperature. What is sought and agreed is optimum function for the occasion rather than optimum temperature *per se*.

It has long been suspected that shivering indicates more than just the misery of exposure to cold. It is in fact a means of generating heat by increasing the rate of muscular activity and so burning more body fuel. Similarly sweating is a means of cooling the body, since the evaporation of even a small amount of water disperses with it a considerable amount of heat. The remarkable discovery, hiding its light under a bushel of commonplace scientific observations on sweating, shivering, and related processes, was that a quantitative assessment of these activities provided a complete and convincing explanation of body temperature regulation. Our ability to

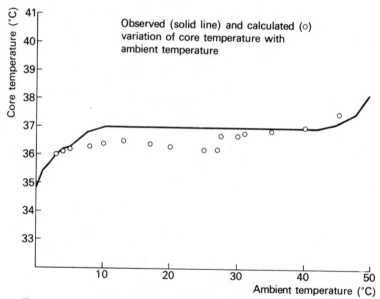

Fig. 4. A comparison between the temperature sustained in the core of a real man (shown as the solid line) with the temperature calculated from the information in Figure 3 (shown as dots). We see that it is possible accurately to account for human temperature regulation by a consensus among the responses of the five separate systems.

sweat or shiver, to burn food or fat, and to control the rate of blood flow to our skin and limbs is all part of a co-operative system for the regulation of our core temperature over an environmental range from freezing to 105° F (40.5°C).

Different animals use each of these regulating processes to a different extent. The dog uses its tongue as the main area for evaporative cooling, as anyone who has seen the winner of a greyhound derby in closeup on television immediately after the event will readily confirm. In addition, human and other animals intentionally seek a warmer or cooler environment, as the case may be, in their ceaseless pursuit of the goal of maximum comfort. If necessary, the local environment is modified to reduce exposure to bearable limits. We wear clothes and build houses; other animals grow fur or seek and make burrows. These activities constitute an additional mechanism of temperature control, which is vital when conditions pass beyond the capacity of internal regulation.

Let us turn for a moment to the philosophical aspect of the subject and consider the problem of pain and discomfort. Some of us are so conditioned to regard unendurable heat, cold, or pain of any kind as in some measure a punishment or visitation from on high for sins of omission or commission that we are inclined to forget that these sensations are all essential components of our survival kit. If shivering and cold were not unpleasant we would not be discussing them, since our remote ancestors would have died of hypothermia. If it seems that such a comment is trite, it is worth considering that C. S. Lewis found it sufficiently serious to be the subject of his book *The Problem of Pain*. It is usual to regard pain as a punishment rather than as a normal physiological phenomenon.

The distinguished American physiologist Walter B. Cannon has said: 'The co-ordinated physiological processes which maintain most of the steady states in the organism are so complex and so peculiar to living things, involving, as they may, the brain and the nerves, the heart, lungs, kidneys and spleen, all working together co-operatively, that I have suggested a special designation for these states, homeostasis.' We shall do well to bear these words in mind when seeking to discover whether there is indeed a process for regulating the planetary temperature, and to look for the exploitation by Gaia of a set of temperature control mechanisms, rather than some simple single means of regulation.

Biological systems are inherently complex, but it is now possible to understand and interpret them in terms of present-day engineering cybernetics, which has advanced far beyond the theory behind the still primitive engineering contrivances used for domestic temperature regulation. Perhaps, in response to our need to conserve energy, we shall eventually devise engineering systems as subtle and flexible as their biological counterparts. The home heat controller may learn to restrict its output to that section of the house where people happen to be, switching parts of itself on and off without human intervention.

To come back to Gaia, how do we recognize an automatic control system when we encounter one? Do we look for the power supply, the regulatory device, or for some complex set of contrivances? As already pointed out, analysis of its parts is usually of little help in showing how a cybernetic system works; unless we know what to look for, recognition of automatic systems by using analytical methods is likely to be just as unsuccessful, whether the system is on a domestic or global scale.

Even though we may find evidence for a Gaian system of temperature regulation, the disentangling of its constituent loops is unlikely to be easy if they are entwined as deeply as in the bodily regulation of temperature. Just as important for Gaia and for all living systems is the regulation of chemical composition. Salinity control, for example, may be a key Gaian regulatory function. If its details are as intricate and complex as those of that amazing organ the kidney, then our quest will be a long one. We now know that the kidney, like the brain, is an information processing organ. To achieve its aim of regulating the salinity of our blood, it purposefully segregates individual atoms. In every second it recognizes and selects or rejects countless billions of atomic ions. This recent new knowledge was not easily found and it may be even more difficult to unravel a system for the global regulation of salinity and chemostasis.

Even a simple control system such as an oven can achieve its purpose in a variety of ways. Imagine an intelligent alien with absolutely no experience of our technology during its last two hundred years of development. He would soon learn to use and recognize a gas oven, but what would he make of one in which food was heated by microwaves?

There is a general approach used by cyberneticians for the recognition of control systems. It is called the black box method, and derives from the teaching of electrical engineering. A student is asked to describe the function of a black box, from which a few wires protrude, without opening the box. He is allowed only to connect instruments or power supplies to the wires and he must then deduce from his observations just what the box is all about.

In cybernetics the black box, or its equivalent, is assumed to be functioning normally. If it is like an oven, it is switched on and cooking. If a living creature, it is alive and conscious. We then test it by changing some property of the environment which we suspect can be controlled by the system we are looking at. If, for example, we are studying human systems and have a co-operative subject, we can take the floor through various angles and at varying rates of speed to discover how well he can stand upright when this fundamental part of his environment is undergoing change. From a simple experiment of this kind we could learn a great deal about the subject's capacity to control his balance. Similarly, with the oven, we could try varying the environmental temperature by using it first in a cold store and then in a hot chamber. We could then observe the limits of external variation consistent with the oven's ability to hold its internal temperature constant. We might also observe the change in power requirements during these environmental transpositions.

This approach to an understanding of control systems, by perturbations of the properties which they are believed to be able to control, is obviously a general one. It can and should always be a gentle one, which if properly conducted in no way impairs the performance or capacity of the system being investigated. The development of this perturbation approach has been somewhat like the evolution of our approach to the study of other living creatures. Not long ago we would kill and dissect them *in situ*. Later, it was recognized that it was better to bring them back alive and look at them in zoos. Nowadays we prefer to watch and observe them in their natural habitats. This more enlightened approach is, alas, not yet general. It may be used in environmental research but in agriculture all too frequently we may leave the animals alone but destroy their habitats, not as a planned perturbation but simply to satisfy our own real or imaginary needs. Many are revolted by the bloody consequences of the hunter's gun or the foxhound's teeth; yet these otherwise sensitive and compassionate people often show little or no concern over the piecemeal death and dispossession wrought by the bulldozer, the plough, and the flame-thrower, in destroying the habitats of our partners in Gaia.

So normal among us all is the acceptance of genocide whilst rejecting murder, the straining at gnats while swallowing camels, that we may well ask ourselves whether this double standard of behaviour is, as altruism is said to be, paradoxically an evolved characteristic favouring the survival of our own kind.

Thus far we have considered cybernetics and control theory only in very general terms. It is beyond the scope of this book to express the cybernetic concepts in the true language of science, mathematics, from which alone can come a complete and quantitative understanding; but we can and must go a little more deeply into this branch of science, which most effectively describes the complex activity of all living things.

Engineers might well be called applied cyberneticians. They use mathematical notation to convey their ideas, together with a few key words and phrases which serve to label the more important concepts of control theory. These descriptive terms are down-to-earth and succinct, and since there is as yet no better way of conveying their meaning in words, we shall now attempt to define them. So let us re-examine our electric oven from an engineer's viewpoint, since the working description provides a convenient and natural context for explaining such cybernetic terms as 'negative feedback'. We have a box made of steel and glass and surrounded by a packing of glass wool or similar material, which serves as a blanket to prevent heat escaping too rapidly and also ensures that the outer surface of the oven is not too hot to touch. Inside, lining the oven walls, are electric heaters. The oven also contains a suitably sited thermostatic device. In the simple oven described earlier, this was a crude affair, no more than a switch designed to turn off the electricity as soon as the desired temperature was reached. The oven we are now studying is a better model, designed for laboratory rather than kitchen use. Instead of an off-and-on switch to control temperature, it has a temperature sensor. This device produces a signal which is proportional to the oven temperature. The signal is in fact an electric current strong enough to activate a temperature gauge, but far too weak to have any heating effect on the oven. In essence it is a device which conveys information rather than power.

The weak signal from this temperature sensor is led to a device which amplifies in much the same way as the amplifier of a radio or television receiver, until it is an electric current powerful enough to heat the oven. The amplifier does not generate electricity; it merely draws on the supply and subtracts a fractional amount from the total requirement to cover its own running costs. Since the signal from the temperature sensor increases in direct proportion to the oven temperature, it cannot be connected directly to the amplifier. If it were, we would have assembled not a temperature-controlled oven but the elements of a cybernetic disaster, and an example of what engineers call 'positive feedback'. As the temperature of the oven rose, the power supplied to the heating elements would increase all the more. A vicious circle would be established and the oven temperature would rise ever more swiftly until the interior became a miniature inferno, or until some cut-out device such as the fuse in the electricity supply broke the circuit.

The correct way in which to join the temperature sensor to the amplifier or, as the engineers would say, 'close the loop', is so that the greater the signal from the temperature sensor, the less the power from the amplifier. This form of connection or loop-closing is called 'negative feedback'. In the oven which we are considering, positive and negative feedback are determined by no more than the order of the two wires from the temperature sensor.

The rapid build-up to disaster in positive feedback, or the precision of temperature control in negative feedback, depends on a property of the amplifier called 'gain'. This is the number of times the weak signal from the sensor is multiplied so as either to enhance or oppose the energy flowing to the heater. Where several loops coexist, each has its own amplifier whose capacity is called the 'loop gain'. In many complicated systems like our own bodies, positive and negative feedback loops coexist. It is obviously helpful to use positive feedback at times, perhaps to restore normal temperature rapidly after a sudden chilling, before negative feedback resumes control.

Grandmother's oven, the kitchen range, where no temperature sensor existed when she was out of the kitchen, is called an 'open loop' device. It would be true to say that the greater part of our search for Gaia is concerned with discovering whether a property of the Earth such as its surface temperature is determined by chance in the open loop fashion, or whether Gaia exists to apply negative and positive feedback with a controlling hand.

It is important to recognize that what is fed back by a sensor is information. This may be transmitted by an electric current, as with our oven, which passes information by varying the strength of its signal. It could equally well be any other information channel, such as speech itself. If, when a passenger in a car, you sense that its speed is hazardous for local conditions and call, 'Too fast: slow down', this is negative feedback. (Assuming the driver heeds your warning. If the wires are unfortunately crossed between you, so that the more you shout slow, the faster he feels impelled to drive, we have another example of positive feedback.)

Information is an inherent and essential part of control systems in another sense, that of memory. They must have the capacity to store, recall, and compare information at any time, so that they may correct errors and never lose sight of their goal. Finally, whether we are considering a simple electric oven, a chain of retail shops monitored by a computer, a sleeping cat, an ecosystem, or Gaia herself, so long as we are considering something which is adaptive, capable of harvesting information and of storing experience and knowledge, then its study is a matter of cybernetics and what is studied can be called a 'system'.

There is a very special attraction about the smooth running of a properly functioning control system. The appeal of the ballet owes much to the graceful and seemingly effortless muscular control of the dancers. The exquisite poise and movements of a 'ballerina assoluta' derive from the subtle and precise interaction of force and counter-force, perfectly timed and balanced. A common failing in human systems is the application of the correcting effort, the negative feedback, too late or too soon. Think of the learner-driver swinging the steering wheel and the car from side to side, through failure to sense in time a drift from his intended course; or think of the drunkard's unsteady progression towards the lamp-post that 'comes out and hits him', as alcohol slows his reactions and he is unable to take avoiding action in time.

Where there is a substantial delay in closing the loop of a feedback system, the correction can turn from negative to positive feedback, especially when events happen within a fairly sharply defined interval of time. The device may then fail by oscillating, sometimes violently, between its limits. Such behaviour can be terrifying when it happens to the steering system of a car, but it is also the source of sound in wind, string, and electronic musical instruments, and of an unceasing array of electronic devices which generate periodic signals of all kinds.

It will now be apparent that the control system of the engineer is one of those forms of protolife mentioned earlier in this book which exist whenever there is a sufficient abundance of free energy. The only difference between non-living and living systems is in the scale of their intricacy, a distinction which fades all the time as the complexity and capacity of automated systems continue to evolve. Whether we have artificial intelligence now or must wait a little longer is open to debate. Meantime we must not forget that, like life itself, cybernetic systems can emerge and evolve by the chance association of events. All that is needed is a sufficient flux of free energy to power the system and an abundance of component parts for its assembly. The level of water in many natural lakes is remarkably independent of the rate of flow in the rivers which feed them. Such lakes are natural inorganic control systems. They exist because the profile of the river which drains the lake is such that a small change in depth leads to a large change in flow rate. Consequently there exists a high-gain negative feedback loop controlling the depth of water in the lake. We must not be misled into assuming that abiological systems of this kind, which might operate on a planetary scale, are purposeful products of Gaia; nor, on the other hand, should we dismiss the possibility of their adaptation and development to serve a Gaian purpose.

This chapter on the stability of complex systems indicates how Gaia may function physiologically. For the present, while the evidence for her existence is still inconclusive, it will serve as one kind of map or circuit diagram to compare with what we may find in further exploration. If we discover sufficient evidence of planet-sized control systems using the active processes of plants and animals as component parts and with the capacity to regulate the climate, the chemical composition, and the topography of the Earth, we can substantiate our hypothesis and formulate a theory.

CHAPTER 5

The contemporary atmosphere

One of the blind spots in human perception has been an obsession with antecedents. Only a hundred years ago that otherwise intelligent and sensitive man, Henry Mayhew, was writing of the poor of London as if they were an alien race. How else could they have been so different from him, he thought. In the Victorian age, almost the same significance was attached to one's family and social background as is now given in some places to one's IQ score. Today, when we hear people extolling breeding and pedigree, they are most likely to be farmers and stock breeders or members of the Jockey or Kennel Clubs.

Yet even now, when interviewing a candidate for a job, we are inclined to attach too much weight to the school and university background and to the academic record. We would rather accept this evidence than take the more difficult step of trying to find out for ourselves what the applicant is really like and what is the potential. Until a few years ago, most of us took a similarly blinkered view of our planet. Attention was focused on its distant past. Textbooks and papers galore were written about the record of the rocks and life in the primaeval seas, and we tended to accept this backward view as telling us all we needed to know about the Earth's properties and potential. It was nearly as bad as trying to assess our job applicant by examining his great-grandfather's bones.

Thanks to what we have learned, and are still learning, about our planet from space research, the whole picture has recently changed. We have had a moon's-eye view of our home in space as it orbits the sun, and we are suddenly aware of being citizens of no mean planet, however mean and squalid the human contribution to this panorama may be in close-up. Whatever happened in the distant past, we are undoubtedly a living part of a strange and beautiful anomaly in our solar system. Our focus has shifted to the Earth we can now study from space, in particular to the properties of its atmosphere. Already we know a great deal more than the most prescient of our forefathers about the composition and behaviour of the insubstantial gaseous veil that encloses the Earth, its denser layers near the surface packed with a curious mixture of reactive gases forever in flux and chemical disarray yet never losing their balance; its tenuous outer filaments clinging with gravitational force to its planetary host and extending a thousand miles into space. However, before we imitate the action of the hydrogen atom and get carried away beyond the atmosphere, let us increase our mass and put a few facts together.

The atmosphere has several well-defined layers. An astronaut travelling upwards from the surface would first pass through the troposphere, the lowest and densest layer. This region of air extends up to about seven miles and is where nearly all the clouds and the weather are. It is also 'the air' for nearly all air-breathing creatures, where there is a direct interaction between the living and gaseous parts of Gaia. It includes more than three-quarters of the total mass. An interesting and unexpected feature of the troposphere, not shared by the other atmospheric layers, is a division into two parts, with the line of separation near the equator. Air from the north and south does not freely mix, as any observer travelling on a ship through tropical regions will readily perceive from the difference in clarity of the skies between the clean southern and the relatively dirty northern hemispheres.

Until very recently it was thought that the gases of the troposphere did not much react among themselves, except perhaps during the fierce heat of a lightning flash or its equivalent. We now know, thanks to the pioneering research in atmospheric chemistry of Sir David Bates, Christian Junge, and Marcel Nicolet, that the gases of the troposphere are reacting just like some planet-sized slow cold flame. Numerous gases are oxidized and scavenged from the air by reactions with oxygen. These reactions are made possible by sunlight which, through a complex chain of events, converts oxygen to more reactive oxygen carriers such as ozone, the hydroxyl radicals, and others.

Above seven to ten miles, depending from where on the Earth's surface he chooses to ascend, our astronaut would enter the stratosphere. This region is so named because the air in it does not easily mix in a vertical direction although fierce winds of hundreds of miles per hour blow at a constant level. The temperature is very low at the lower boundary of the stratosphere, the tropopause, but rises as we travel upwards. The nature of the two layers is intimately associated with the temperature gradients within them. The troposphere, with its constant fall of about 1°C for every hundred metres climbed in height, makes the vertical motion of the air easy and the formation of clouds with their familiar shapes the rule.

In the stratosphere, where it grows warmer aloft, hot air is reluctant to rise and consequently stratified stability is the rule. The shorter and more powerful wavelengths of the sun's ultra-violet rays penetrate the upper stratosphere where they split oxygen into oxygen atoms. These soon combine again, but often to form ozone. The ozone is split also by the ultra-violet rays and so an equilibrium is established with about five parts per million of ozone at maximum density. The air of the stratosphere is not much denser than that of Mars, so no oxygen-breathing life could survive there; indeed, even if the low pressure was overcome by a pressurized environment, life would rapidly be destroyed by ozone poisoning. As the passengers and crew of certain high-flying long-range airliners have recently found out to their peril and discomfort, stratospheric air, even when brought to tolerable temperatures and pressures inside the aircraft, is not fit to breathe. Smog is healthier by comparison.

The chemistry of the stratosphere is of the greatest interest to academic scientists. Countless chemical reactions proceed under the pure abstract conditions of the gas phase. There are no walls as in laboratory vessels to mar the perfection. It is not surprising, therefore, that almost all scientific work on atmospheric chemistry so far has concentrated on the stratosphere and higher regions. It has a special name—chemical aeronomy—chosen by that most famous aeronomist, Sidney Chapman. Yet apart from the postulated but unproven consequences of ozone changes, life at the surface seems less involved with the upper regions than are its representative scientists. These remarks are not made in criticism but in reflection on the fact that science tends to follow what can be measured and discussed. It so happens that the greater part of the atmosphere, the troposphere, has been least measured and understood, yet it is certainly the part most relevant to Gaia.

Above the stratosphere, in the ionosphere, the air is very thin indeed and as we ascend and encounter the fierce unfiltered rays of the sun the pace of chemical reactions accelerates. In these regions most molecular species other than nitrogen and carbon monoxide tend to split into their constituent atoms. Some atoms and molecules are further dissociated into positive ions and electrons, thus forming the electrically conducting layers which, in the days before orbiting man-made satellites, were important for their capacity to reflect radio-waves and allow global communication.

The outermost layer of air, so thin as to contain only a few hundred atoms per cubic centimetre, the exosphere, can be thought of as merging into the equally thin outer atmosphere of the sun. It used to be assumed that the escape of hydrogen atoms from the exosphere gave Earth its oxygen atmosphere. Not only do we now doubt that this process is on a sufficient scale to account for oxygen but we rather suspect that the loss of hydrogen atoms is offset or even counterbalanced by the flux of hydrogen from the sun. Table 3 (p. 68) shows the principal reactive gases of the air, their concentrations, residence times, and main sources of origin.

As explained earlier, I first became interested in the possibility of the terrestrial atmosphere being a biological ensemble, rather than a mere catalogue of gases, when testing the theory that an analysis of the chemical composition of a planetary atmosphere would reveal the presence or absence of life. Our experiments confirmed the theory and at the same time convinced us that the composition of the Earth's atmosphere was so curious and incompatible a mixture that it could not possibly have arisen or persisted by chance. Almost everything about it seemed to violate the rules of equilibrium chemistry, yet amidst apparent disorder relatively constant

Table 3. *Some chemically reactive gases of the air*

Gas	Abundance %	Flux in megatons per year	Extent of disequilibrium	Possible function under the Gaia hypothesis
Nitrogen	79	300	10^{10}	Pressure builder Fire extinguisher Alternative to nitrate in the sea
Oxygen	21	100,000	None. Taken as reference	Energy reference gas
Carbon dioxide	0.03	140,000	10	Photosynthesis Climate control
Methane	10^{-4}	1,000	Infinite	Oxygen regulation Ventilation of the anaerobic zone
Nitrous oxide	10^{-5}	100	10^{13}	Oxygen regulation Ozone regulation
Ammonia	10^{-6}	300	Infinite	pH control Climate control (formerly)
Sulphur gases	10^{-8}	100	Infinite	Transport gases of the sulphur cycle
Methyl chloride	10^{-7}	10	Infinite	Ozone regulation
Methyl iodide	10^{-10}	1	Infinite	Transport of iodine

Note: Infinite in column 4 means beyond limits of computation and favourable conditions for life were somehow maintained. When the unexpected occurs and cannot be explained as an accidental happening, it is worth seeking a rational explanation. We shall see if the Gaia hypothesis accounts for the strange composition of our atmosphere, with its proposition that the biosphere actively maintains and controls the composition of the air around us, so as to provide an optimum environment for terrestrial life. We shall therefore examine the atmosphere in much the same way that a physiologist might examine the contents of the blood, to see what function it serves in maintaining the living creature of which it is a part.

From a chemical viewpoint, although not in terms of abundance, the dominant gas of the air is oxygen. It establishes throughout our planet the reference level of chemical energy which makes it possible, given some combustible material, to light a fire anywhere on the Earth. It provides the chemical potential difference wide enough for birds to fly and for us to run and keep warm in winter; perhaps also to think. The present level of oxygen tension is to the contemporary biosphere what the high-voltage electricity supply is to our twentieth-century way of life. Things can go on without it, but the potentialities are substantially reduced. The comparison is a close one, since it is a convenience of chemistry to express the oxidizing power of an environment in terms of its reduction-oxidation (redox) potential, measured electrically and expressed in volts. It is in fact no more than the voltage of a hypothetical battery with one electrode in the oxygen and the other in the food.

Nearly all the oxygen produced by photosynthesis in green plants and algae is cycled through the atmosphere and used up in that other fundamental activity of life, respiration, in a relatively short space of time. This complementary process can obviously never yield a net increment of oxygen. How then has oxygen accumulated in the atmosphere?

It was thought until recently that the main source was the photolysis of water vapour in the upper layers, where water molecules are split and the hydrogen atoms are light enough to escape the Earth's gravitational field, leaving the oxygen atoms to couple in molecules of gas or to bond triply in ozone. This process certainly produces a net increment of oxygen, but important though it may have been in the past, it is a negligible source of oxygen in the contemporary biosphere. There seems little doubt that the principal source of oxygen in the atmosphere is the one first proposed by Rubey in 1951, namely, the burial in sedimentary rocks of a small proportion of the carbon which is fixed by green plants and algae in the organic matter of their own tissues. Approximately 0.1 per cent of the carbon fixed annually is buried with the plant debris which is washed and blown down from the land surfaces into the seas and rivers, leaving one additional oxygen molecule in the air for each carbon atom thus removed from the cycle of photosynthesis and respiration. Were it not for this process, oxygen would be steadily withdrawn from the air by reaction with reducing materials exposed by weathering, earth movements, and volcanic outgassing.

It is somewhat cynically said that the eminence of a scientist is measured by the length of time that he holds up progress in his field. Among the great scientists, Pasteur was no exception to this rule. He was responsible for the assumption that before oxygen appeared in the air, only low-grade forms of life were possible. This notion has had a long run but, as indicated in chapter 2, we now believe that even the first photosynthesizers operated with as high a chemical potential as is available to micro-organisms today. In the beginning the large potential energy gradient at present provided by oxygen was then only available within the cells of these organisms. Later, as they multiplied, it extended to their micro-environment and continued spreading with life until the primaeval reducing materials of the Earth were all oxidized and oxygen was free at last to appear in the air. From the beginning, however, the potential energy difference between the oxidants of the photosynthesizing cells and the reducing environment outside was as great as that which exists between oxygen outside and food within the cells today.

Sources of high potential, whether chemical or electrical, are dangerous. Oxygen is particularly hazardous. Our present atmosphere, with an oxygen level of 21 per cent, is at the safe upper limit for life. Even a small increase in concentration would greatly add to the danger of fires. The probability of a forest fire being started by a lightning flash increases by 70 per cent for each 1 per cent rise in oxygen concentration above the present level. Above 25 per cent very little of our present land vegetation could survive the raging conflagrations which would destroy tropical rain forests and arctic tundra alike. Andrew Watson of Reading University has recently confirmed experimentally the probability of fire under a range of conditions closely similar to those of natural forests. This is illustrated in the diagram below.

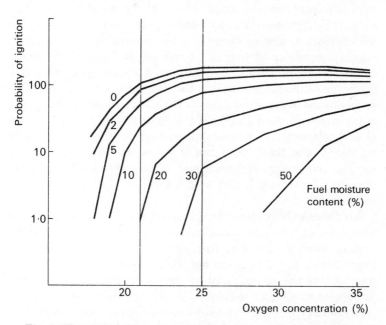

Fig. 5. The probability of grass or forest fires in atmospheres of different oxygen enrichment. Natural fires are started by lightning strokes or by spontaneous combustion; their probability is greatly dependent upon the moisture content of the natural fossil fuels. Each line corresponds to a different moisture level going from completely dry (0%) to visibly wet (45%). At the present oxygen content (21%) fires do not start at more than 15% moisture content. At 25% oxygen even the damp twigs and grass of a rain forest would ignite.

The present oxygen level is at a point where risk and benefit nicely balance. Forest fires do indeed take place, but not with sufficient frequency to interfere with the high productivity that a 21 per cent oxygen level permits. This again is just like electricity. Energy losses in transit and the amount of copper required in cables are greatly reduced as the supply voltage is raised, but a 250-volt supply for domestic consumption is about as high as can be justified without the risk of death by shock or fire becoming unacceptable.

The power-station engineers do not let their equipment run haphazardly. It is designed and operated with great care and skill to ensure that the supply of electricity to our homes is at a constant safe potential. How, then, is the oxygen level in the air controlled? Before discussing the nature of this biological regulation, we need to look at the composition of the atmosphere in more detail. To examine a single gas through a telescope or microscope or in a test-tube tells us little about its relationship to the other gases of the air. It is rather like trying to comprehend the meaning of a sentence from an examination of a single word. The information content of the atmosphere resides in the total ensemble of gases, so we must consider oxygen, our energy reference gas, in relation to other gases of the air with which it can and does react. Let us start with methane.

Hutchinson first showed us, thirty years ago, that methane, or marsh gas, was a biological product. He thought it came mostly from the farts of ruminant animals. Although their contribution is not to be denied, we now know that the greater part of this gas is produced by bacterial fermentation in the anaerobic muds and sediments of the sea beds, marshes, wet lands, and river estuaries where carbon burial takes place. The quantity of methane made in this way by micro-organisms is astonishingly large, at least 1,000 million tons a year. (The 'natural' gas pumped into our homes comes from a different stable; this is fossil gas—the gaseous equivalent of coal or oil, and the supply is trivial on a planetary scale. In a decade or so, the small reservoirs of 'natural' gas will be exhausted.)

Within the context of a self-regulating biosphere actively maintaining its gaseous environment at an optimum for life, it is appropriate to ask what is the function of a gas such as methane. It is no more illogical than asking what is the function of glucose or of insulin in the blood. In a non-Gaian context, the question would be condemned as circular and meaningless, which may be why it has not been asked long before.

What, then, is the purpose of methane and how does it relate to oxygen? One obvious function is to maintain the integrity of the anaerobic zones of its origin. As methane bubbles up continuously through those fetid muds, it sweeps them free of poisonous volatile substances such as the methyl derivatives of arsenic and lead, and also of course, from the viewpoint of the anaerobes, of that poisonous element oxygen itself.

When methane reaches the atmosphere it appears to act as a two-way regulator of oxygen, capable of taking at one level and putting a little back at another. Some of it travels to the stratosphere before oxidizing to carbon dioxide and water vapour, thus becoming the principal source of water vapour in the upper air. The water ultimately dissociates to oxygen and hydrogen. Oxygen descends and hydrogen escapes into space. By this means a small but possibly significant addition of oxygen to the air is ensured in the long term. When the books are balanced, an escape of hydrogen always means a net gain of oxygen.

Conversely, the oxidation of methane in the lower atmosphere uses up substantial amounts of oxygen, of the order of 2,000 megatons annually. This process goes on slowly and continuously in the air we live and move in, through a series of complex and subtle reactions unravelled largely through the work of Michael McElroy and his colleagues. Simple arithmetic shows that in the absence of methane production, the oxygen concentration would rise by as much as 1 per cent in as little as 12,000 years: a very dangerous change and, on the geological time-scale, a far too rapid one.

Rubey's theory of oxygen balance, as developed by Holland and Broecker and other eminent scientists, proposes that the abundance of oxygen in the atmosphere is kept constant by a balance between a net gain as carbon is buried and a net loss by the reoxidation of reduced materials expelled from below the Earth's crust. However, the biosphere is too powerful an engine to be left to run with what the engineers call a passive control system, as if in a power station the boiler pressure was determined by a balance between the quantity of fuel burnt and the quantity of steam required to drive the turbines. On warm Sundays when little power was required, pressure would rise until the boiler would be in danger of exploding, while at peak demand periods the pressure would fall and consumption could not be met. For this reason engineers use active control systems. As explained in chapter 4, these have a sensing element such as a pressure gauge or a thermometer which detects any departure from the optimum requirements and uses a little of the system's power supply to alter the rate at which fuel is burnt.

The constancy of oxygen concentration suggests the presence of an active control system, presumably with a means of sensing and signalling any departure from the optimum oxygen concentration in the air; this may be linked with the processes of methane production and carbon burial. Once carbonaceous matter has reached the deep anaerobic zones, it is destined either to make methane or to be buried. At present, nearly twenty times as much carbon is used to produce 1,000 megatons of methane annually as is buried. Hence any mechanism which can alter this proportion will effectively regulate oxygen. Perhaps when there is too much oxygen in the air, some warning signal may be amplified in the course of methane production, and steady-state conditions could then be promptly restored by the upwelling of this regulatory gas into the atmosphere. The energy apparently wasted in methane oxidation is now seen to be the inevitable power requirement of an active, short-time constant regulator. It is an intriguing thought that without the assistance of those anaerobic micro-flora living in the stinking muds of the sea-beds, lakes, and ponds, there might be no writing or reading of books. Without the methane they produce, oxygen would rise inexorably in concentration to a level at which any fire would be a holocaust and land life, apart from micro-flora in damp places, would be impossible.

Another puzzling atmospheric gas is nitrous oxide. Like

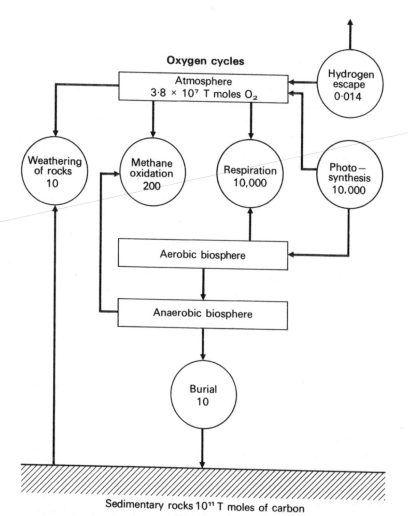

Fig. 6. The fluxes of oxygen and carbon between the major reservoirs of the Earth's atmosphere, surface, and oceans. Quantities are shown in units of terramoles. A terramole of carbon is 12 megatons and of oxygen 32 megatons. Figures inside the circles are annual fluxes. Figures in the two reservoirs, the atmosphere and the sedimentary rocks, indicate their size. Note how the carbon, en route for burial in the sedimentary layers beneath the sea and the marshes and wet lands, is for the most part vented back to the atmosphere as 'marsh gas', methane.

methane, it is present at near to one-third part per million in the air and, again like methane, its minute concentration bears no relation to the scale of its production rate by the micro-organisms of the soil and of the seas. Nitrous oxide is made at the rate of between 100 and 300 megatons a year. This is approximately the rate at which nitrogen itself is returned to the air. There is an abundance of nitrogen and very little nitrous oxide around because nitrogen is a very stable gas and accumulates, whereas nitrous oxide is rapidly destroyed by the ultra-violet rays of the sun.

We may be sure that the efficient biosphere is unlikely to squander the energy required in making this odd gas unless it has some useful function. Two possible uses come to mind, and since it is a commonplace in biology for the same material to serve more than one end, both may be important. First, nitrous oxide may be involved, like methane, in the task of oxygen regulation. The quantity of oxygen carried to the atmosphere by nitrous oxide from the soil and sea beds is twice the amount needed to balance the losses due to the oxidation of reducing materials which are constantly exposed from below the Earth's surface. Nitrous oxide may therefore be a possible counterbalance to methane. It is at least conceivable that methane and nitrous oxide production are complementary and may be yet another means of promptly regulating oxygen concentration.

The second possibly important activity of nitrous oxide concerns its behaviour in the stratosphere where it decomposes to give, among other products, nitric oxide, which has been claimed to be catalytically destructive of ozone. This may seem an alarming situation in view of the warnings of many environmentalists that the worst catastrophe at present threatening our world is the destruction of the ozone layer in the stratosphere by supersonic transport or the products of aerosol spray-cans. In fact, if oxides of nitrogen do deplete ozone, nature has been in the business of destroying the ozone layer for a long, long time. Too much ozone may be as bad as too little. Like everything else in the atmosphere, there are desirable optima. The ozone layer could increase possibly by as much as 15 per cent. For all we know, more ozone might be climatically undesirable. We certainly do know that ultra-violet light from the sun is useful and beneficial in some respects, and a thicker ozone layer might prevent it from reaching the Earth's surface in sufficient quantity. In humans vitamin D is formed by exposure of the skin to ultra-violet radiation. Too much ultra-violet may mean skin cancer; too little most certainly means rickets. Although we cannot expect some particular global beneficence for us as a species from the micro-organisms through their production of nitrous oxide, low-level ultra-violet may be valuable to other species in ways we do not yet know. A regulating device would at least seem helpful and nitrous oxide, together with that other recently discovered atmospheric gas of biological origin, methyl chloride, may serve this purpose. If so, the Gaian control system will include a means of sensing whether too much or too little ultra-violet radiation is getting through the ozone layer, and of regulating nitrous oxide production accordingly.

Another nitrogenous gas made in large volumes in the soil and the sea and released to the air is ammonia. It is a difficult gas to measure, but its estimated rate of production is not less than 1,000 megatons a year. As with methane, the biosphere uses a great deal of energy in producing ammonia, which is now entirely of biological origin. Its function is almost certainly to control the acidity of the environment. When the total production of acids by the oxidation of nitrogen and sulphur is taken into account, the ammonia produced by the biosphere is found to be just sufficient to sustain a rainfall pH near 8, the optimum for life. In the absence of ammonia the rain everywhere would fall at a pH close to 3, which is about as acid as vinegar. In parts of Scandinavia and North America this is already happening and is said to be drastically reducing growth. It is thought to be caused by the burning of industrial and domestic fuels both in the affected areas and other heavily populated regions around. Most fuel contains sulphur and after combustion much of it returns to the ground as sulphuric acid borne by rain droplets and carried to the stricken areas by the prevailing winds.

Life can be tolerant of acidity. The digestive juices in our stomachs are proof of that, but an environment as acid as vinegar is far from optimal. It is indeed fortunate that almost everywhere in the natural world, ammonia and acids are in balance and the rain is neither too acid nor too alkaline. If we assume that this balance is actively maintained by Gaia's cybernetic control system, the energy cost in terms of ammonia production will be charged against the total photosynthetic account.

By far the most abundant constituent of the atmosphere is nitrogen gas which makes up 79 per cent of the air we breathe. The bonds uniting two nitrogen atoms to form a molecule of nitrogen gas are among the strongest in chemistry and it is therefore reluctant to react with anything. It has accumulated in the atmosphere because denitrifying bacteria and other processes of living cells have put it there. It is only slowly returned to its natural habitat, the sea, by inorganic processes such as thunderstorms.

Few people are aware that the stable form of nitrogen is not the gas but the nitrate ion dissolved in the sea. As we saw in chapter 3, if life were deleted most of the nitrogen of the air would eventually combine with oxygen and return to the sea in nitrate form. What advantages are there to the biosphere in keeping the air pumped up with nitrogen gas, far from the expectations of equilibrium chemistry? There are several possibilities. Firstly, a stable climate may require the present atmospheric density and nitrogen is a convenient pressure-building gas. Secondly, a slow reacting gas like nitrogen is probably the best diluent for oxygen in the air and, as we have seen, a pure oxygen atmosphere would be disastrous. Thirdly, if all the nitrogen were in the sea in the form of nitrate ion, it would add to the always tricky problem of keeping the salinity low enough for life. As we shall see in the following chapter, a cell membrane is extremely vulnerable to the salt level of its environment and is destroyed by a total salinity in excess of 0.8 molarity. It makes no difference whether the salt is chloride or nitrate or a mixture of the two. If all nitrogen were in the sea in nitrate form, the molarity would be increased from 0.6 to 0.8. This would raise the total ionic strength of sea water to a level inconsistent with almost all known forms of life. A final point is that high concentrations of nitrate are toxic quite apart from their effect on the salinity of the sea.

Adaptation to an environment with a high level of nitrate might have been more difficult and energy-consuming for the biosphere than simply storing nitrogen away in the air, where it could do some good. Any of these would seem a valid reason for the biological processes which return nitrogen from the sea and the land to the air.

It is obvious that the abundance of an atmospheric gas is no measure of its importance. For example, ammonia is one hundred million times less abundant than nitrogen, yet its role may be just as important from the point of view of regulation. Indeed, the amount of ammonia produced annually is as great as that of nitrogen, but the turnover of ammonia is much faster. The abundance of the gases of the air is determined far more by their rate of reaction than by their rate of production. The rarer gases tend to be major participants in the business of life.

One of the most valuable contributions to contemporary chemistry has been the unravelling of the intricate chemical reactions of the atmospheric gases. We now realize, for example, that trace gases such as hydrogen and carbon monoxide are intermediate products in the reaction between methane and oxygen, and can therefore be considered as biological gases like their progenitors. Many other trace reactive gases of the air such as ozone, nitric oxide, and nitrogen dioxide fall into this category, together with a great number of very transient reaction substances which the chemists call free radicals. One of these is the methyl radical, the first product of the oxidation of methane. About 1,000 million tons of it pass annually through the air but because the life-span of the methyl radical is less than one second, its abundance may be no more than one for each cubic centimetre of air. This is not the place in which to give a full account of the complex chemistry of these reactive radicals, but it is an interesting story for those who would like to know more about the gases of the air.

The so-called rare and noble gases of the air are neither particularly rare nor entirely noble. At one time they were believed to be resistant to attack by any chemical agent; in other words, like those noble metals, gold and platinum, they would pass the acid test. We now know that two of them, krypton and xenon, can both form compounds. The most abundant gas of this family is argon which, together with helium and neon, makes up nearly 1 per cent of the air and so can hardly be called rare. These inert gases of unequivocally inorganic origin are useful in helping us to establish more clearly the lifeless background, like our perfectly flat sandy beach, against which life is revealed.

The man-made gases such as the fluorocarbons, which have their sources mainly in the chemical industry and were never in the air before industrial man appeared, are very indicative of life at work. A visitor viewing the Earth from outer space and discovering aerosol-propellant gases in our atmosphere, would have no doubt whatever that our planet bore life, and probably intelligence of a kind as well. In our persistent self-imposed alienation from nature, we tend to think that our industrial products are not 'natural'. In fact, they are just as natural as all the other chemicals of the Earth, for they have been made by us, who surely are living creatures. They may of course be aggressive and dangerous, like nerve gases, but no more so than the toxin manufactured by the *botulinus* bacillus.

Finally we come to those essential components of the atmosphere, and of life itself, carbon dioxide and water vapour. Their importance for life is fundamental but it is difficult to establish the possibility that they may be biologically regulated. Most geochemists agree that the carbon dioxide content of the atmosphere, 0.03 per cent, is kept constant in the short term by simple reactions with sea water. For the technically-minded, carbon dioxide and water are in equilibrium with bicarbonic acid and its anion in solution.

Nearly fifty times as much carbon dioxide is loosely held in the ocean in this form as there is in the air. If the carbon dioxide content of the air should fall for any reason, the normal level would be restored by the release of some of the enormous reserve held by the ocean. At the present time the amount of carbon dioxide in the atmosphere is increasing because of the widespread consumption of fossil fuels. If we stopped burning these fuels tomorrow, it would not take long, perhaps thirty years, for atmospheric carbon dioxide to revert to its normal level, as equilibrium between the quantity of gas in the air and the bicarbonate in the sea was re-established. Fossil fuel combustion has in fact increased the amount of carbon dioxide in the air by about 12 per cent. The significance of this man-made change in the atmosphere is discussed in chapter 7.

If Gaia regulates carbon dioxide, it is probably by the indirect method of assisting the attainment of equilibrium rather than by opposing it. To return to our analogy of the sandy beach, it would be the purposeful smoothing of an area of lumpy sand before starting to build a sand-castle. However, it is not easy to distinguish between induced and natural states of equilibrium, and it may be a case of deciding on circumstantial evidence alone.

On the long-term geological time-scale the equilibrium between the silicate rocks and carbonate rocks of the ocean floor and of the Earth's crust, as proposed by Urey, should provide even larger reserves to ensure a constant level of carbon dioxide. With the situation so well in hand, is there any need for Gaia's intervention? There may certainly be a need, if the attainment of equilibrium is not rapid enough for the biosphere as a whole. It is like a man finding one spring morning that he cannot leave his house for work because the door is blocked by snow. He knows that it will melt away in time, but he cannot afford to wait for nature to take its course and must clear it quickly with a shovel.

There are many signs of Gaian impatience with the leisurely progress towards natural equilibrium in the case of carbon dioxide. Most life forms contain the enzyme carbonic anhydrase which speeds the reaction between carbon dioxide and water; a constant rain of carbonate-bearing shells sinks towards the ocean floor, where it ultimately forms beds of chalk or limestone rock and thus prevents the stagnation of carbon dioxide in the upper layers of the sea; and Dr A. E. Ringwood has suggested that the ceaseless break-up of the soil and rocks by all forms of life speeds the reaction between carbon dioxide, water, and carbonate rocks.

It seems possible that without life's interference, carbon dioxide would accumulate in the air until dangerous levels might be reached. As a 'greenhouse' gas, its presence with water vapour in the contemporary atmosphere keeps the temperature tens of degrees above what it would otherwise be. If, due to fossil fuel combustion, the level of carbon dioxide rose too rapidly for inorganic equilibrium forces to cope, the threat of overheating might become serious. Fortunately, this green-house gas interacts strongly with the biosphere. Not only is carbon dioxide the source of carbon for photosynthesis but it is also removed from the atmosphere and converted into organic matter by many heterotrophic (that is, non-photosynthetic) organisms. Even animals incorporate a little atmospheric carbon dioxide and it is, of course, given off in respiration by nearly all organisms. In fact, the more it seems that inorganic equilibrium or steady-state processes determine the atmo-spheric concentration of a gas, the greater may be the extent of its biological involvement. This is not surprising in the context of a biosphere which is actively controlling its environment and whose policy is always to turn existing conditions to its own advantage.

Biological involvement with that strange and versatile chemical substance hydrogen oxide, otherwise known as water, follows a similar pattern but is even more fundamental. The cycling of water from the oceans through the atmosphere to the land surfaces is powered largely by solar energy, yet life insists on participating through the process of transpiration. Sunlight may distil water from the sea, later to fall as rain on the land, but sunlight does not spontaneously at the Earth's surface split oxygen from water and drive reactions leading to the synthesis of intricate compounds and structures.

Earth is the water-planet. Without water there would have been no life, and life is still utterly dependent on its impartial generosity. It is the ultimate background of reference. All departures from equilibrium might be considered as depar-tures from the water-reference level. The properties of acidity and alkalinity, and of oxidizing and reducing potentials are estimated in relation to the neutrality of water. The human species uses mean sea level as the base point against which heights and depths are measured.

Like carbon dioxide, water vapour has the properties of a greenhouse gas and interacts strongly with the biosphere. If we accept the proposition that life actively controls and adapts the atmospheric environment to its needs, its relationship with water vapour illustrates our conclusion that the incompatibilities of biological cycles and inorganic equilibria are more apparent than real.

CHAPTER 6

The sea

As Arthur C. Clarke has observed: 'How inappropriate to call this planet Earth, when clearly it is Ocean.' Nearly three-quarters of the Earth's surface is sea, which is why those magnificent photographs taken from space show our planet as a sapphire blue globe, flecked with soft wisps of cloud and capped by brilliant white fields of polar ice. The beauty of our home contrasts sharply with the drab uniformity of our lifeless neighbours, Mars and Venus, which both lack that abundant covering of water.

The oceans, those great expanses of deep blue sea, have far more to them than the mere capacity to dazzle an observer in outer space. They are vital parts of the global steam engine that transforms the radiant energy from the sun into the motions of air and water which in turn distribute this energy over all regions of the world. Collectively, the oceans form a reservoir of dissolved gases which helps to regulate the composition of the air we breathe and to provide a stable environment for marine life—about half of all living matter.

We are not sure how the oceans were formed. It was so long ago, well before the start of life, that almost no tangible geological evidence has persisted. There have been many hypotheses about the form of the early oceans, even that the Earth was once entirely covered by ocean without any land or even any shallow water. The land and continents came later. If this hypothesis is ever established, then those about the origins of life will need to be revised. However, there is still general agreement that the oceans derived from the Earth's interior some time after it had accreted as a planet and had warmed up sufficiently to distil off the gases and water of the primaeval atmosphere and seas.

The history of the Earth before life is not directly helpful in our quest for Gaia. More relevant and interesting is the physical and chemical stability of the oceans ever since life appeared. There is evidence that over the past three and a half aeons, while the continents formed and drifted about the globe, the polar ice melted and refroze, and the sea level rose and fell, the total volume of water has remained unchanged, despite the metamorphoses. At present, the average depth of the oceans is 3,200 metres (about 2 miles), although there are trenches as deep as 10,000 metres (about 6 miles). The total volume of water is around 1.2 thousand million cubic kilometres (300 million cubic miles) and its weight is about 1.3 million million million tonnes.

These are big numbers which must be put in perspective. Although the weight of the oceans is 250 times that of the atmosphere, it only represents 1 part in 4,000 of the weight of the Earth. If we were to model the Earth by a globe 30 centimetres in diameter, the average depth of the sea would be little more than the thickness of the paper on which these words are printed, while the deepest trench would be represented by a dent of one-third of a millimetre.

Oceanography, the scientific study of the sea, is usually reckoned to have started about a hundred years ago with the voyage of the research ship *Challenger*. This ship carried out the first systematic study of all the oceans of the world. Its programme included observations of the physics, chemistry, and biology of the sea. In spite of this promising multi-disciplinary start, oceanography has since fragmented into separate sub-sciences: marine biology, chemical oceanography, ocean geophysics, and other hybrid subjects, of which there are as many as there are professors to defend them as territories. Yet for all this it has been a comparatively neglected science. Most work of significance has been done since the Second World War, spurred on by international competition for fresh sources of food, power, and strategic advantage generally. A return to the spirit of the *Challenger* expedition, with its concept of the sea as indivisible, is at last beginning to gain ground. The physics, chemistry, and biology of the oceans are once again being recognized as interdependent parts of a huge global process.

A practical starting-point in our search for Gaia in the oceans is to ask ourselves why the sea is salt. The answer that was once confidently given (and no doubt still appears in many a standard text and encyclopaedia) ran something like this: the sea became salt because rain and rivers constantly washed small amounts of salt from the land into the sea. The surface waters of the oceans may evaporate and later fall on the land as rain, but salt, a non-volatile substance, is always left behind and accumulates in the sea. The oceans therefore become more saline as time goes by.

This answer certainly fits in neatly with the traditional explanation of why the salt content of the body fluids of living creatures, including ourselves, is lower than that of the oceans. At present, the salt content of the sea expressed as a percentage (the number of parts by weight of salt in one hundred parts of water) is about 3.4 per cent, while that of our blood is about 0.8 per cent. The explanation goes thus: when life began, the internal fluids of marine organisms were in equilibrium with the sea, or in other words the salinity of the organic fluids and the salinity of their environment were exactly equal. Later, when life took one of its evolutionary leaps forward and sent migrants from the seas to colonize the land, the internal salinity of living organisms became fossilized, as it were, at the prevailing level, while that of the sea continued to increase. Hence the difference today between the salinity of organic fluids and of the sea.

This theory of salt accumulation, if correct, should enable us to calculate the age of the oceans. There is no difficulty in estimating their total content of salt at the present time, and if we assume that the quantity of salt washed into the sea every year by the action of rain and rivers has remained more or less the same throughout the ages, a simple division should give the correct answer. The input of salt into the sea is about 540 megatons each year; the total volume of sea water is 1.2 thousand million cubic kilometres; the average salinity is 3.4 per cent. Therefore the time taken to reach the present level of salinity is about 80 million years, which must be the age of the oceans. However, this answer is manifestly out of step with all of palaeontology. So let us start again.

Ferren MacIntyre has recently pointed out that continental run-off is not the only source of salt in the sea. He recalls an old Norse myth that the sea is salt because somewhere at the bottom of the ocean there is a salt-mill forever grinding away. The Norsemen were not far wrong because we now know that the plastic doughy rocks of the Earth's hot interior at times well up and push through the ocean floor, which consequently spreads. This process, which is part of the mechanism that causes the continents to drift apart, also adds more salt to the sea. When the salt from this source is added to the amount washed off from the land and our calculation is repeated, the age of the oceans becomes 60 million years. In the seventeenth century the Irish Protestant divine, Archbishop Ussher, calculated the age of the Earth from the chronology of the Old Testament. His figures gave the date of Creation as 4004 BC. He was wrong, but in comparison with the true time-scale, his figures seem scarcely further out than the estimate of 60 million years for the age of the oceans.

It seems reasonably certain that life began in the sea and geologists have produced evidence of the existence of simple organisms, probably bacteria, nearly three and a half aeons ago. The oceans must be at least as old. This is consistent with estimates of the age of the Earth obtained from radioactive measurements, indicating that it was formed about four and a half aeons, or 4,500 million years ago. Geological evidence also shows that the salt content of the sea has not in fact changed very much since the oceans came into being and life began; not enough, in any event, to account for the difference between the present salinity of the sea and that of our blood.

We are forced by discrepancies of this kind to re-think the whole problem of why the sea is salt. The figures for the rate of addition of salt to the oceans through continental run-off (rain and rivers) and sea-floor spreading (the 'salt-mill') are reasonably firm, yet the salinity level has not increased by anything like the amount to be expected from the salt-accumulation theory. The only possible conclusion seems to be that there must be a 'sink' through which salt disappears from the oceans at the same rate as it is added. Before we speculate about the nature of this sink and what happens to the salt flowing through it, we need to consider some aspects of the physics, chemistry, and biology of the sea.

Sea-water is a complex but thin soup of living and dead organisms and of dissolved or suspended inorganic compounds. The main dissolved constituents are the inorganic salts. In the language of chemistry, the word 'salt' describes a class of compounds of which sodium chloride, common salt, is only one example. The composition of sea-water varies from place to place across the globe and with depth below the surface. In terms of total salinity the variations are small, although of much importance in the detailed interpretation of oceanic processes. However, for our present purpose, which is to discuss the general mechanism of salt control, we shall disregard these variations.

The average sample of sea-water contains 3.4 per cent per kilogram weight of inorganic salts, of which about 90 per cent is sodium chloride. This statement is not strictly accurate in scientific terms, because when inorganic salts dissolve in water they split into two separate sets of atom-sized particles with opposite electric charges. These particles are known as ions. Thus sodium chloride splits into a positive sodium and a negative chloride ion. In solution, the two kinds of ion wander more or less independently among the surrounding water molecules. This may seem surprising, since opposite electric charges attract one another, and normally stay together in ion-pairs. The reason why, when in solution, they do not is that water has the property of weakening by a large factor the electrical forces between oppositely charged ions. If the solutions of two different salts are mixed together, for example sodium chloride and magnesium sulphate, all one can say about the composition of the resulting solution is that it is a mixture of four ions: sodium, magnesium, chloride, and sulphate. In suitable conditions one can in fact separate sodium sulphate and magnesium chloride from the mixture more readily than the original salts of sodium chloride and magnesium sulphate.

Strictly speaking, it is therefore incorrect to say that sea-water 'contains' sodium chloride; it contains the constituent ions of sodium chloride. It also contains magnesium and sulphate ions together with much smaller amounts of other ionic components such as calcium, bicarbonate, and phosphate, which play an indispensable role in the living processes that take place in the sea.

One of the lesser-known requirements for a living cell is that, with rare exceptions, the salinity of either its internal fluids or its external environment must never exceed for more than a few seconds a value of 6 per cent. A few creatures can live in brine pools and salt lakes with salinity levels above this limit, but they are as exceptional and bizarre as those micro-organisms that can survive in boiling water. Their special adaptations have been made by permission of the rest of the living world, which provides oxygen and food in suitable form and ensures that these essentials are transferred to the brine pool or hot spring. Without this help, these strange creatures could not survive, in spite of their remarkable adaptations to their near-lethal habitats.

Brine shrimps, for example, possess an extraordinary tough shell which is as impermeable to water as is the hull of a submarine. This enables them to retain the same internal salinity as we have—around 1 per cent—while living in very salt water. Without the protection of this tough shell, these creatures would dry out and shrivel up in a matter of seconds as the water of the mildly salt solution inside them flowed out to dilute the stronger salt solution of the brine pool.

This tendency of water to move from a weaker to a stronger saline solution is an example of what physical chemists call osmosis. Osmosis takes place whenever a solution of low salt concentration—or any other dissolved material, for that matter—is separated from a solution of higher concentration by a wall that lets through the water but not the salt. The water flows from the weaker to the stronger solution so that the higher concentration is diluted. Other things being equal, the process continues until the two solutions are in equilibrium.

The flow can be halted by applying a mechanical force to oppose it. The opposing force is called osmotic pressure and its operation depends on the nature of the dissolved material and on the difference between the strengths of the two solutions. Osmotic pressure may be considerable. If the casing round the brine shrimp let water through, the pressure that the shrimp would have to exert to prevent itself from being dehydrated would be about 150 kilograms per square centimetre, or 2,300 pounds per square inch, which is equal to the pressure exerted by a column of water a mile high. Or we might say that if the shrimp had to obtain the water it needs for its internal workings from the salt lake, making water flow from a stronger to a weaker solution, it would need to have inside it a pump capable of taking water from a well one mile deep.

Osmotic pressure is thus a consequence of differences between internal and external salinities. Provided that both concentrations are below the critical level of 6 per cent, most living organisms can deal quite easily with the engineering problem involved. The absolute level is what matters, for faced with an internal or external salinity above 6 per cent, living cells literally fall to pieces.

Living processes consist largely of interactions between macro-molecules. A meticulously programmed sequence of events usually takes place, in which two large molecules, for example, may approach one another, position themselves accurately, stay close together for a while, during which time material may be exchanged, and then separate. Accurate positioning is achieved with the help of electric charges variously sited on each macro-molecule. The positively charged areas on one molecule mesh precisely with the negatively charged areas on the other. With living systems, these interactions invariably take place in a water environment, where the presence of dissolved ions modifies the natural electrical attraction of macro-molecules and enables them to approach one another and position themselves with due deliberation and a high degree of precision.

In effect, positive ions cluster around the negative areas of macro-molecules and negative ions around the positive areas. The ion cluster acts as a kind of screen that partly neutralizes the charge it is surrounding and thereby reduces the attraction of one macro-molecule towards another. The higher the salt concentration, the greater will be the screening effect of the ions and the weaker will be the forces of attraction. If the concentration is too high, the macro-molecules may cease to interact and that part of the cell function will fail. If the salt concentration is too low, the attraction forces between adjacent macro-molecules will become irresistible, the molecules will fail to separate, and the orderly sequence of reactions will suffer disruption of another kind.

The material composing the surface membrane of a living cell is held together by electrical forces similar to those involved in macro-molecular processes. This enclosing membrane ensures that the salinity of the internal contents of the cell remains within permissible limits. Hardly more substantial than a soap film, it is as effective a barrier to leakage of the cell's constituents as the hull of a ship to water or the fuselage of an aircraft to the outside atmosphere. However, the water-tightness of a living cell is achieved by quite different means from that of a ship's hull. The latter works mechanically and statically; the cell wall does its job by active and dynamic use of biochemical processes.

The thin film that encapsulates every living cell contains ion-pumps which selectively exchange internal for external ions according to the cell's needs. Electrical forces ensure that the cell-membrane has the flexibility and strength required for these operations. If the concentration of salt on either side of the membrane exceeds the critical level of 6 per cent, the screening effect of the salt-derived ions clustered around the electric charges holding the membrane together is too strong. The tension is lost, the weakened membrane disintegrates, and the cell falls apart. Except for the highly specialized mem-branes of the halophilic (salt-loving) bacteria of the brine pools, the membranes of all living creatures are subject to this salinity limit.

We can see now why living organisms, deeply dependent on the operation of electrical forces, can survive only if the salinity of the environment is held within safe limits, and particularly within the critical upper limit of 6 per cent. In the light of this knowledge, we begin to lose interest in the original question: 'Why is the sea salt?' Continental run-off and sea-floor spreading easily account for the present level of salinity in the oceans. The more important question is: 'Why isn't the sea *more* salt?' Catching a glimpse of Gaia, I would answer:

'Because since life began, the salinity of the oceans has been under biological control.' The next question is obviously: 'But how?' Which brings us to the crux of the matter, for what we really need to know and think about is not how salt is added to the sea but how it is removed. We are in fact back at the sink looking for a salt-removal process which must be in some way linked with the biology of the sea if our belief in the intervention of Gaia is well-founded.

Let us restate the problem. There is comparatively reliable evidence, both direct and indirect, that the salinity of sea-water has changed very little in hundreds of millions, if not thousands of millions, of years. Our knowledge of the level of salinity tolerated by living organisms of the type that have thrived in the sea over these vast periods suggests that in any event the salinity cannot have exceeded 6 per cent, compared with the present level of 3.4 per cent, and that even if it had risen as high as 4 per cent, life in the sea would have evolved through quite different organisms from those revealed by the fossil record. Yet the amount of salt washed off the land into the sea by rain and rivers during each 80-million-year period is equal to the present total salt content of the oceans. If this process had continued unchecked since their formation, all the oceans would now be like the Dead Sea, saturated with salt and an intolerable habitat for life.

A means must therefore exist for the removal of salt from the sea as fast as it is added. The need for such a device has long been recognized by oceanographers, and several attempts have been made to identify it. The various theories put forward have all relied essentially on non-living inorganic mechanisms, but none has found general acceptance. Broecker has stated that the way in which sodium and magnesium salts are withdrawn from the sea is one of the great unsolved mysteries of chemical oceanography. In fact, two problems need to be solved, for the removal of the positive sodium and magnesium ions and of the negative chloride and sulphate ions have to be treated separately, since positive and negative ions exist independently in a watery medium. To complicate matters further, more sodium and magnesium than chloride and sulphate ions are added to the sea by continental run-off, and to keep things electrically stable, the positive charge carried by the excess ions of sodium and magnesium has to be balanced by negatively charged ions of aluminium and silicon.

Broecker has tentatively suggested that sodium and magnesium are removed by being dropped in the rain of debris that falls constantly to the sea bed, thus becoming part of the sediment, or that they somehow combine with the minerals that constitute the ocean floor. Unfortunately no independent evidence to support either possibility has so far been obtained.

An entirely different mechanism is needed to account for the removal and disposal of the negative chloride and sulphate ions. Broecker points out that water will evaporate from isolated arms of the sea, such as the Persian Gulf, more rapidly than it can enter from rivers or rain. If evaporation is prolonged, salts will crystallize out in vast deposits, which will eventually be overlaid and buried by natural geological processes. These great beds of salt are to be found below ground and beneath the continental shelves all over the world, and at the surface also.

The time scale of these processes is of the order of hundreds of millions of years and is therefore consistent with the salinity record—except in one vital respect. If we assume that the formation of isolated arms of the sea and the upheavals of the Earth's crust which lead to the burial of salt beds are due entirely to inorganic processes, we must also accept that they will occur entirely at random, both in time and space. They may account for the *average* level of salinity of the oceans staying within tolerable limits, but large and lethal fluctuations would inevitably have occurred as a result of the random nature of the control processes.

It is surely time for us to ask ourselves whether the presence of the living matter with which the seas abound could have modified the course of events and may still be acting to solve this difficult problem. Let us start by reviewing the possible living components of the mechanism which would enable such engineering feats on a planetary scale to be performed.

About half of the mass of the living matter in the world is to be found in the sea. Life on the land is for the most part two-dimensional, held by gravity to the solid surface. Marine organisms and the sea have about the same density; life is freed from the limitations of gravity, and the pasture is three-dimensional. The primary living forms that capture the energy of the sun and turn it into food and oxygen by the process known as photosynthesis, thus energizing the entire ocean, are freely floating cells, in contrast to the ground-anchored photosynthesizing plants on land. Trees are neither found nor needed in the sea and there are no grazing herbivores, but only large grazing carnivores—the whales which feed by sweeping up the myriads of minute shrimp-like crustaceans known as krill.

The sequence of life in the sea begins with the primary producers, those countless millions of single-celled, free-floating microscopic plants, or micro-flora, which biologists call phytoplankton. These provide pasture for the microscopic animals known as zooplankton, which are preyed on by larger creatures, and so on, through a series of carnivores of ever-increasing size and rarity. The sea, unlike the land, is therefore numerically dominated by tiny single-celled *protista*, including algae and protozoa. These thrive only in the sunlit, upper, 100-metre layer of the oceans. Of especial note are the coccolithopores, with shells of calcium carbonate which often contain a drop of oil to serve as a combined aquatic balloon and food store, and the diatoms, a type of algae with skeletal walls made of silica. These and many other kinds form part of the complex and diverse flora of what is called the euphotic zone.

It is worth looking in more detail at the role of the diatoms in the oceans. Diatoms and their close relatives, the radiolarians, are particularly beautiful. Their skeletons are made of opal, fashioned in a variety of intricate and always exquisite designs. Opal is a special gem-like form of silicon dioxide, usually known as silica, the main constituent of sand and quartz. Silicon is one of the most abundant elements in the Earth's crust; most rocks, from clay to basalt, contain it in combined form. It is not generally considered of importance in biology—there is little silicon in us or in anything we eat—but it is a key element in the life of the sea.

Broecker discovered that less than 1 per cent of the silica-bearing minerals washed into the sea from the land is retained

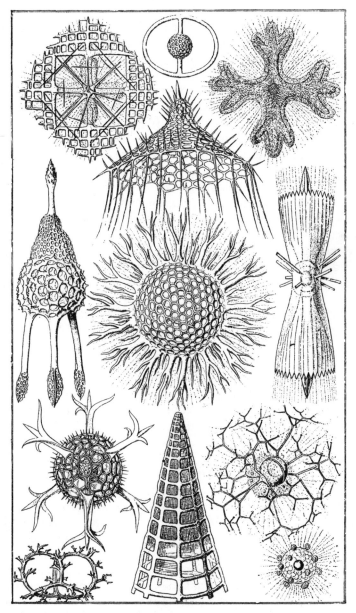

Fig. 7. Deep-sea radiolaria from the *Challenger* expedition.
From Haeckel, *History of Creation*, vol. 2.

in the surface waters. On the other hand the ratio of silica to salt in dead, landlocked salt lakes is much higher than in the sea, as one might expect in a lifeless environment where conditions are close to chemical equilibrium. Diatoms, which assimilate silica and flourish in the sea but obviously not in the salt-saturated lakes, spend their brief lives in surface waters. When they die, they sink to the ocean floor and their opaline skeletons pile up as sediment, adding about 300 million tons of silica to the sedimentary rocks each year. Thus the life cycle of these microscopic organisms accounts for the silicon deficiency in the surface waters of the sea, and contributes to its vigorous departure from chemical equilibrium.

This biological process for the use and disposal of silica can be seen as an efficient mechanism for controlling its level in the sea. If, for example, increasing amounts of silica were being washed into the sea from the rivers, the diatom population would expand (provided that sufficient nitrate and sulphate nutrients were also in good supply) and reduce the dissolved silica level. If this level fell below normal requirements, the diatom population would contract until the silica content of the surface waters had built up again, and this is well known to occur.

We can now ask ourselves whether this silica control mechanism follows Gaia's general pattern for controlling the constituents of sea-water, notably sea-salt. Is this how life intervenes to solve the problems inherent in Broecker's theories of purely inorganic machinery for the control and disposal of sea-salt?

From a planetary engineering point of view, the significance of the life cycle of diatoms and coccoliths is that when they die their soft parts dissolve and their intricate skeletons or shells sink to the bottom of the sea. A constant rain of these structures, which oceanographers call 'tests', almost as beautiful in death as in life, has fallen on the ocean floor through the aeons, building up great beds of chalk and limestone (from coccoliths) and silicate (from diatoms). This deluge of dead organisms is not so much a funeral procession as a conveyor belt constructed by Gaia to convey parts from the production zone at surface levels to the storage regions below the seas and continents. Some of the soft organic matter falls all the way with the inorganic skeletons and ultimately turns into buried fossil fuels, sulphide ores, and even free sulphur. The whole process has the advantage of built-in yet flexible control systems, based on the responsiveness of living organisms to changes in their environment and their capacity to restore, or adapt to, conditions which favour their own survival.

Now for some proposals about possible Gaian devices for controlling salinity. Although still conjecture, I believe these ideas are solid enough to serve as bases for detailed theoretical and experimental study.

Let us start with a possible means of speeding up the ocean conveyor-belt system. It is probable that salts are carried down into the sediment by being entrapped in the descending rain of plant and animal debris, as Broecker suggested, just as ordinary rain entraps atmospheric dust particles. It may be that there are species of hard-shelled marine *protista* or animals that are particularly sensitive to salinity and die off rapidly as soon as the level rises even fractionally above normal. Their shells fall, carrying salt with them to the sea bed and thus reducing its effective level in surface waters. The quantities of salt which could be sequestered from the ocean by this process are too small to account directly for the sink we seek. However, as we shall see, a link between the rate of deposition of tests and salt levels could be part of a method for regulating the sea's salinity.

Another and quite different possibility arises from Broecker's proposal for removing chloride and sulphate. He suggested that excess salt accumulates in the form of evaporites in shallow bays, landlocked lagoons, and isolated arms of the sea, where the rate of evaporation is rapid and the inflow from the sea is one-way. Let us now make the bold speculation that the lagoons formed as a consequence of the presence of life in the sea. If this process then progressed to homoeostasis, it could solve Broecker's problem of accounting for the stability of a salt-removal system apparently based on the formation of evaporites by entirely random inorganic forces.

Constructing vast barriers of the size needed to enclose thousands of square miles of sea in tropical regions may seem an engineering task well beyond human capabilities. Yet larger by far than any man-made structures are the coral reefs, and more significantly in former times the stromatolite reefs. These are constructed on a Gaian scale, with city walls miles high and thousands of miles long, built by a co-operative of living organisms. Is it possible that the Great Barrier Reef, off the north-east coast of Australia, is the partly finished project for an evaporation lagoon?

Even if it has no Gaian significance, this example of what fairly simple living creatures have achieved by co-operation over the aeons encourages us to speculate on other possibilities. We have already seen how the atmosphere has been changed world-wide by living creatures. What are we to make of volcanic activity and continental drift? Both are consequences of the inner motions of our planet, but could Gaia also be at work? If so, would they not provide additional mechanisms for lagoon building, quite apart from their primary effect on sea-floor spreading and the transfer of sediments?

Speculations of this kind are by no means as far-fetched as they might at first appear. Oceanographers already suspect that under-sea volcanoes may sometimes be the end results of biological activity. The connection is quite straightforward. Much of the sediment falling on the ocean floor is almost pure silica. In time, the pressure of this accumulating deposit on the thin plastic rock of the sea floor is heavy enough to dimple it, and more weight of sediment sinks into the depression. Meantime, the conduction of heat from the Earth's interior is impeded by this ever-thickening silica blanket, whose open structure makes it a good heat insulator in the manner of a woollen blanket; the temperature in the zone under the sediment increases; the underlying rock softens even more, the layer deforms, another thickness of sediment is laid down in the hollow and the temperature rises still higher. These are the conditions of positive feedback. Eventually, the heat is intense enough to melt the rock of the sea floor and volcanic lava pours out. Volcanic islands could be formed in this manner; perhaps sometimes lagoons as well. In the shallower waters near the coast vast deposits of calcium carbonate are laid down. Sometimes these emerge again as chalk or lime-

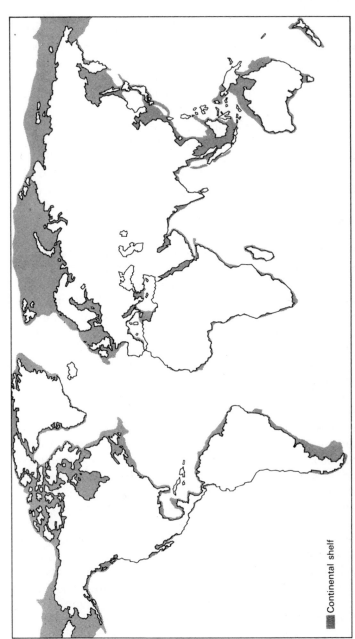

Fig. 8. The continental shelves of the oceans. These regions, which occupy an area as large as the African continent, may be crucial in the homoeostasis of our planet. Here carbon is buried which sustains oxygen in the air, and here is the source of many other gaseous and volatile compounds essential for life.

Continental shelf

stone. At other times they are drawn down into the hot rocks beneath, where they act as a flux to melt the rock and so assist in the construction of volcanoes.

In a lifeless sea, the sediment needed to set off this chain of events might never have lodged itself in the right place. Volcanoes are found on dead planets but, judging by the large one on Mars known as Nix Olympus, they are very unlike their terrestrial counterparts. If Gaia has modified the sea floor, it has been done by exploiting a natural tendency and turning it to her own advantage. I am not, of course, suggesting that all or even most volcanoes are caused by biological activity; but that we should consider the possibility that the tendency towards eruptions is exploited by the biota for their collective needs.

If the idea of major upheavals of the Earth's crust being manipulated in the interests of the biosphere still offends common sense, it is worth reminding ourselves that man-made dams have occasionally started earthquakes by altering the weight distribution over the surrounding area. The disturbance potential of a mass of sediment or a coral reef is infinitely greater.

Our discussion of salinity and its control is incomplete and very general. I have said practically nothing about variations in salinity from place to place in oceanic waters; nothing about such saline components as phosphate and nitrate ions, primary nutrients whose relationships are still a puzzle to oceanographers; nothing about the manganese nodules found over wide areas of the sea bed, whose origin is undoubtedly biological; and nothing about the complexities of oceanic currents and circulation systems. All are processes, or parts of processes, which directly or indirectly influence, or are influenced by, the presence of living matter. I have hardly touched on the question of the ecological relationships among thousands of species of marine organisms; on whether man's intentional or chance interference with their way of life could have repercussions on the physics and chemistry of the oceans and subsequently on our own wellbeing; whether, for example, the slaughter of whales, which could lead to the complete extinction of these wonderful mammals, might have other far-reaching effects, apart from depriving us forever of their unique company. These omissions are partly due to lack of space, but even more to a lack of solid information on which to build.

Fortunately, steps are at last being taken to fill the many empty shelves of our information store. Expenditure on a 'Big Science' scale is not always necessary. A few years ago some of us took part in a modest project aimed at studying some of the specialist but nonetheless important activities of Gaia, which are on a somewhat smaller scale than the large engineering works on which we have speculated in relation to salinity control.

In 1971 I set sail with two colleagues, Robert Maggs and Roger Wade, in a small research vessel, the *Shackleton*, of only a few hundred tons displacement, on a voyage from Barry in South Wales to Antarctica, the main purpose of which was to carry out geological surveys. We three were supernumeraries, free to use the ship as a moving observation platform while it sailed towards the south and the fulfilment of its mission. Our special object was to investigate the possibility of balancing the world's sulphur budget by including a previously unsuspected but potentially important component, dimethyl sulphide.

The mystery of the sulphur gap began a few years earlier, when scientists tracing the sulphur cycle found that more sulphur was regularly washed off the land by rivers into the sea than could be derived from all the known sources on land. They took into account the weathering of sulphur-bearing rocks, the sulphur extracted from the ground by plants, and the amounts put into the air by the burning of fossil fuels; yet they were still faced with a discrepancy of the order of hundreds of millions of tonnes each year. E. J. Conway, pursuing the idea that the missing sulphur component was probably carried from the sea to the land via the atmosphere, suggested that it was hydrogen sulphide, that malodorous gas which gave old-fashioned school chemistry the name of 'Stinks'. Our party doubted this simple explanation. For one thing, neither we nor anyone else had ever found enough hydrogen sulphide in the atmosphere to account for the size of

the discrepancy; for another, hydrogen sulphide reacts so rapidly with the oxygen-rich sea-water, forming non-volatile products, that it could never have time to reach the sea's surface, or to escape from it into the air. My colleagues and I favoured instead the compound dimethyl sulphide, a relative of hydrogen sulphide in the chemical sense, as the agent that carried the missing sulphur through the air. It had one essential qualification for the role—its rate of destruction by oxygen was very much slower than that of the rival candidate, hydrogen sulphide.

We had good grounds for backing dimethyl sulphide. Professor Frederick Challenger of Leeds University had demonstrated over many years of experimentation that adding methyl groups (the process known as methylation) to certain chemical elements was a device frequently adopted by organisms to rid themselves of an excess of unwanted substances by turning them into gases or vapours. The methyl compounds of sulphur, mercury, antimony, and arsenic, for example, are all much more volatile than the elements themselves. Challenger had shown that many species of marine algae, including the seaweeds, were able to produce dimethyl sulphide in this manner, in large quantities.

We sampled sea-water throughout the voyage and found dimethyl sulphide to be present in what then seemed to us sufficient strength to qualify it for the role of the sulphur bearer. However, Peter Liss later convinced us by calculation that the strength of our mid-ocean samples indicated that there could not be enough dimethyl sulphide in the sea to build up and sustain the large airward flow required to account for all the missing sulphur. We realized later still that the *Shackleton*'s passage did not bring us into contact with the marine waters where dimethyl sulphide is produced in strength. The principal source of this substance, as we have since discovered, is not the deep ocean—relatively speaking a desert—but inshore waters rich in living matter. This is where one finds certain kinds of algal seaweed with an astonishingly efficient mechanism for extracting sulphur from sulphate ions in the sea and converting it to dimethyl sulphide. One of these algae is *Polysiphonia fastigiata*, a small red organism which lives attached to the large bladder-wrack to be seen on most sea shores. It is so productive of dimethyl sulphide that if it is placed in a closed jar partly filled with sea-water and left for about thirty minutes, enough dimethyl sulphide builds up to make the vapour in the air space almost flammable. Happily, the smell of dimethyl sulphide is nothing like that of hydrogen sulphide. In dilute form its pleasant odour is redolent of the sea.

Although more investigation is needed to prove our case, it now seems reasonable to propose that dimethyl sulphide produced in the sea around the continental shelves is the missing sulphur bearer. Many algal species can assume both salt and fresh-water forms. The Japanese scientist Ishida has recently shown that both forms of *Polysiphonia fastigiata* are capable of producing dimethyl sulphide, but the effective enzyme system is only switched on in sea-water. This may suggest a biological device for ensuring the production of dimethyl sulphide in the right place for feeding into the sulphur cycle.

There is a grim side to the process of biological methylation. The bacteria living in the mud of the sea bed have developed this technique extensively; toxic elements such as mercury, lead, and arsenic are all converted here to their volatile methyl forms. These gases are carried up through the sea-water and permeate everything, including the fish. In normal circumstances the quantities are too small to be poisonous, but some years ago Japanese industry on the shores of the inland Sea of Japan actually discharged dimethyl mercury into the sea, so that its strength in the marine environment was enough to make the fish poisonous to man. All who ate the fish suffered, many were painfully crippled, and some died of Mimamata disease, the local name for the unique and horrifying nature of methyl mercury poisoning. It is fortunate that the natural process of mercury methylation does not proceed to this drastic extent. Not so with arsenic. In the last century some wall-papers were coloured by a green pigment made from arsenic. In damp and mouldy houses with poor ventilation the mould turned the arsenic in the wall-papers to a lethal gas, trimethyl arsine, and the sleepers in the bedrooms so decorated died.

The biological object of methylating poisonous elements is

not known for certain, but it seems likely that it is a means of removing toxic substances from the local environment by converting them into gaseous form. Dilution normally prevents these gases from harming other creatures, but when man disturbs the natural balance this beneficial process becomes malign, with crippling or lethal consequences.

The biological methylation of sulphur appears to be Gaia's way of ensuring a proper balance between the sulphur in the sea and on the land. Without this process, much of the soluble sulphur on land surfaces would have been washed off into the sea long ago and never replaced, thus disturbing the delicate equations between the environmental constituents needed for the maintenance of living organisms.

Another group of methyl-containing compounds claimed our attention during the *Shackleton* voyage. These were the so-called 'halocarbons', substances derived from hydrocarbons such as methane by replacing one or more of the hydrogen atoms with one of the elements fluorine, chlorine, bromine, or iodine. Chemists know this group of elements collectively as the halogens. This investigation was to prove the most positive scientific contribution of our voyage, and a typical example of how in basic exploratory research it is unwise to plan ahead in too fine detail; one must keep one's eyes open and see what Gaia has to offer. By good fortune we had taken with us a piece of equipment that could be used to measure minute traces of halocarbon gases. Our prime intention was to see if the release of aerosol-propellant gases, such as are used to spray deodorants and insecticides, would effectively label the air and enable us to observe its motions between, for example, the Northern and Southern Hemispheres. This research was in some respects only too successful. Wherever we sailed we found it easy to detect and measure the fluorochlorocarbon gases, and this discovery led directly to the present possibly exaggerated concern over their capacity to deplete the ozone layer.

Our apparatus also revealed the presence of two more halocarbon gases. One was carbon tetrachloride, whose presence in the air is still an enigma; the other was methyl iodide, a product of a marine alga.

Some of us may recall those long strips of seaweed which

used to be hung up to foretell the weather. They are a form of kelp or, to use the botanical name, laminaria. They grow in inshore waters and have the capacity to gather iodine from the sea. During growth, they produce methyl iodide in prolific quantities. Kelp used to be harvested and burnt, iodine being extracted from the ashes. It seems probable that just as dimethyl sulphide acts as a sulphur bearer, methyl iodide conveys back to the land through the air that life-essential element, iodine. Without iodine, the thyroid gland could not produce the hormones which regulate the metabolic rate, and most animals would eventually sicken and die.

At the time when we discovered methyl iodide in the air over the sea, we were unaware that most of this gas reacts with the chloride ions of the sea to produce methyl chloride. Oliver Zafiriou first drew our attention to this unexpected reaction and we are indebted to him, for it led to the discovery of methyl chloride as the principal chlorine-bearing gas of the atmosphere. In the ordinary way it would have been little more than a chemical curiosity but, as indicated in the previous chapter, methyl chloride is now seen to be the natural equivalent of the aerosol-propellant gases, in its capacity to deplete the ozone layer of the stratosphere. It may serve to regulate the density of the ozone layer and is a reminder that too much ozone can be as harmful as too little. So yet another element, chlorine from the sea, with methyl attached, is a candidate for a Gaian role.

Other elements important to life, such as selenium, may also be found to pass from the sea to the air as methyl derivatives, but so far we have failed to find a marine source of a volatile compound of that key element, phosphorus. It is possible that the needs for phosphorus are small enough to be met by the weathering of the rocks, but in case this should not be so, it is worth asking ourselves whether the movements of migratory birds and fish serve the larger Gaian purpose of phosphorus recycling. The strenuous and seemingly perverse efforts of salmon and eels to penetrate inland to places distant from the sea would then be seen to have their proper function.

Gathering information about the sea, its chemistry, physics, and biology and their interacting mechanisms, should come right at the top of mankind's list of priorities. The more we know, the better we shall understand how far we can safely go in availing ourselves of the sea's resources, and the consequences of abusing our present powers as a dominant species and recklessly plundering or exploiting its most fruitful regions. Less than a third of the Earth's surface is land. This may be why the biosphere has been able to contend with the radical transformations wrought by agriculture and animal husbandry, and will probably continue to strike a balance as our numbers grow and farming becomes ever more intensive. We should not, however, assume that the sea, and especially the arable regions of the continental shelves, can be farmed with the same impunity. Indeed, no one knows what risks are run when we disturb this key area of the biosphere. That is why I believe that our best and most rewarding course is to sail with Gaia in view, to remind us throughout the voyage and in all our explorations that the sea is a vital part of her.

CHAPTER 7

Gaia and Man: the problem of pollution

Nearly all of us have been told more than once by our tribal elders that things were better in the good old days. So ingrained is this habit of thought—which we pass on in our turn as we grow old—that it is almost automatic to assume that early man was in total harmony with the rest of Gaia. Perhaps we were indeed expelled from the Garden of Eden and perhaps the ritual is symbolically repeated in the mind of each generation.

Biblical teaching that the Fall was from a state of blissful innocence into the sorrowful world of the flesh and the devil, through the sin of disobedience, is hard to accept in our contemporary culture. Nowadays it is more fashionable to attribute our fall from grace to man's insatiable curiosity and his irresistible urge to experiment and interfere with the natural order of things. Significantly, both the biblical story and, to a lesser extent, its modern interpretation seem aimed at inculcating and sustaining a sense of guilt—a powerful but arbitrary negative feedback in human society.

Perhaps the first thing that comes to mind about modern man which might justify the belief that he is still hell-bent is the increasing pollution of the atmosphere and the natural waters of our planet since the Industrial Revolution, which began in Britain in the late eighteenth century and spread like a stain through most of the Northern Hemisphere. It is now generally accepted that man's industrial activities are fouling the nest and pose a threat to the total life of the planet which grows more ominous every year. Here, however, I part company with conventional thought. It may be that the white-hot rash of our technology will in the end prove destructive and painful for our own species, but the evidence for accepting that industrial activities either at their present level or in the immediate future may endanger the life of Gaia as a whole, is very weak indeed.

It is all too easily overlooked that Nature, apart from being red in tooth and claw, does not hesitate to use chemical warfare if more conventional weapons prove inadequate. How many of us recognize that the insecticide which is sprayed in the home to kill flies and wasps is a product of chrysanthemums? Natural pyrethrum is still one of the most effective substances for killing insects.

By far the most poisonous substances known are natural products. The *Botulinus* toxin produced by bacteria, or the deadly product of the algal dinoflagellates which cause the red tide at sea, or the polypeptide manufactured by the death-cap fungus: all three are entirely organic products and but for their toxicity would be suitable candidates for the shelves of the health food store. The African plant *Dichapetalum toxicarium* and some related species have learnt to perform fluorine chemistry. They incorporate the fiery element fluorine within natural substances such as acetic acid, and fill their leaves with the resulting salt compound. This deadly substance has been referred to by biochemists as a metabolic monkey-wrench, which graphically illustrates the havoc it causes at a molecular level when drawn into the gear wheels of the chemical cycles of almost any other living organism. If it were solely an industrial product, it would be cited as yet another example of man's perverse and wicked use of chemical technology to hit below the belt and improve his position in the biosphere. Yet it is a natural product and only one of many highly toxic substances which are made organically and enable their possessors to take a mean advantage. There is no Geneva Convention to limit natural dirty tricks. One of the *Aspergillus* family of moulds has discovered how to make a substance called aflatoxin which is mutagenic, carcinogenic, and teratogenic; in other words it can cause mutations, tumours, and foetal abnormalities. It is known to have caused a vast amount of human misery through stomach cancer resulting from the eating of peanuts naturally polluted by this aggressive chemical.

Could it be that pollution is natural? If by pollution we mean the dumping of waste matter there is indeed ample evidence that pollution is as natural to Gaia as is breathing to ourselves and most other animals. I have already referred to the greatest pollution disaster ever to affect our planet, which took place one and a half aeons ago with the emergence of free oxygen gas in the atmosphere. When this happened, all the Earth's surface in contact with the air and with the tidal seas became lethal for a large range of micro-organisms. These, the anaerobes (organisms which can only grow in the absence of oxygen), were in consequence driven to an underground existence in the muds at the bottom of rivers, lakes, and sea beds. Many millions of years later, their banishment from life at the top was to some extent revoked. They are now back again on the surface in the most comfortable and secure of environments, enjoying a truly pampered existence and optimum status, while continuously supplied with food. These minute organisms now inhabit the gut of all animals from insects to elephants. My colleague, Lynn Margulis, believes that they represent one of the more important aspects of Gaia, and it may well be that large mammals including ourselves serve mainly to provide them with their anaerobic environment. Although this affair, the widespread destruction of the anaerobes, eventually had a happy ending, in no way does it minimize the extent of the oxygen pollution disaster at the time when it happened. To illustrate the effect of oxygen poisoning on the anaerobic life, I have already postulated a marine alga able to generate chlorine by photosynthesis, which successfully took over the oceans.

The natural disaster of oxygen pollution occurred slowly enough to allow the ecosystems of that time to adapt, although it must have been at the expense of numerous species until a new ecosystem made up of those resistant to oxygen inherited the surface of the Earth.

The relatively minor environmental upheaval occasioned by the Industrial Revolution illustrates how such adaptations can take place. There is the well-known example of the Peppered Moth which in a few decades changed its wing colour from pale grey to nearly black so as to preserve its camouflage against predators as it rested on the soot-covered trees of England's industrial areas. It is now fast changing back to grey again as the impact of the Clean Air Act eliminates the soot. But roses still bloom better in London than in remote country areas, a consequence of the destruction by the pollutant sulphur dioxide of the fungi which attack them.

The very concept of pollution is anthropocentric and it may even be irrelevant in the Gaian context. Many so-called pollutants are naturally present and it becomes exceedingly difficult to know at what level the appellation 'pollutant' may be justified. Carbon monoxide, for example, which is poisonous to us and to most large mammals, is a product of incomplete combustion, a toxic agent from exhaust gases of cars, coke or coal-burning stoves, and cigarettes; a pollutant put into otherwise clean fresh air by man, you might think. However, if the air is analysed we find that carbon monoxide gas is to be found everywhere. It comes from the oxidation of methane gas in the atmosphere itself and as much as 1,000 million tons of it are so produced each year. It is thus an indirect but natural vegetable product and is also found in the swim-bladders of many sea creatures. The syphonophores, for example, are loaded with this gas in concentrations which would speedily kill us off if present in our own atmosphere at similar levels.

Almost every pollutant, whether it be in the form of sulphur dioxide, dimethyl mercury, the halocarbons, mutagenic and carcinogenic substances, or radioactive material, has to some extent, large or small, a natural background. It may even be produced so abundantly in nature as to be poisonous or lethal from the start. To live in caves of uranium-bearing rock would be unhealthy for any living creature, but such caves are rare enough to present no real threat to the survival of a species. It seems that as a species we can already withstand the normal range of exposure to the numerous hazards of our environment. If for any reason one or more of these hazards should increase, both individual and species adaptation will set in. The normal defensive response of an individual to an increase in ultra-violet light, for example, is the bronzing of the skin. In a few generations this becomes a permanent . change. The fair-skinned and freckled do not flourish if exposed to the tropical sun, but the species suffers only if racial taboos prevent free access for future generations to the native genes which confer pigmentation.

When a species, through some accident of genetic chemistry, inadvertently produces a poisonous substance, it may well kill itself. However, if the poison is more deadly to its competitors it may manage to survive and in time both adapt to its own toxicity and produce even more lethal forms of pollutant, as Darwinian selection runs its course.

Let us now examine contemporary pollution from a Gaian rather than a human angle. So far as industrial pollution is concerned, by far the most heavily affected places are the densely populated urban areas of the north temperate zones: Japan, parts of the USA, of Western Europe and of Soviet Russia. Many of us have had a chance to view these regions from the window of an aeroplane in flight. Provided that there is enough wind to disperse the smog so that the surface is visible, the usual sight is of a green carpet lightly speckled with grey. Industrial complexes stand out, together with the close-packed housing of the workers, yet the general impression is that everywhere the natural vegetation is biding its time and waiting for some unguarded moment that will give it a chance to return and take over everything again. Some of us remember the rapid colonization by wild flowers of city areas cleared by bombing in the Second World War. Industrial regions seldom appear from above to be the denatured deserts which the professional doomsters have led us to expect. If this is true of the most polluted and populous areas of our planet, it may seem that there is no urgent cause for concern about man's activities. Unfortunately this is not necessarily so; it is merely that we have been led to look for trouble in the wrong places.

Influential people, the opinion-shapers and law-makers in all societies, tend to live or at least work in cities and to travel to and from their work by roads and railways threading through corridors of urban and industrial development. Their daily journeys make them depressingly aware of urban pollution and of local environments which are rarely pleasant to pass through or gaze at in a traffic hold-up. Holidays in less developed regions by the sea or among mountains confirm by contrast the belief that their home or working base is unfit for life. It also strengthens their determination to do something about it.

Thus the understandable but erroneous impression has been created that the greatest ecological disturbance is in the urbanized regions of the temperate zone of the Northern Hemisphere. A flight over the Harappan Desert of Pakistan or over many parts of Africa, or not so long ago over the south-central areas of the United States, the background of Steinbeck's novel *The Grapes of Wrath*, would have given a more accurate and revealing picture of devastation in both the natural and man-made ecosystems. It is in these regions of vast disturbance, the dust bowls, that man and his livestock have most markedly lowered the potential for life. These disasters were not caused by the over-enthusiastic use of advanced technology; on the contrary, it is now generally acknowledged that they were the fruits of unsound and bad husbandry, supported by a primitive technology.

It is instructive to compare these devastations with the contemporary English scene. Here, agricultural productivity, helped considerably by the resources of industry, has grown to such an extent that the country now produces more food than is needed to support life, in spite of a population density of over 1,000 per square mile, one of the highest in the world. There is still space to spare for gardens, parks, woodlands, wastelands, hedgerows and coppices, to say nothing of towns, roads, and industry. It is true that in his enthusiasm for increased profit and productivity the farmer has tended to use his industrial machinery more in the manner of a butcher than a surgeon, and he is still inclined to regard all living things other than his livestock and crops as pests, weeds, or vermin. Yet this could be a passing phase in the renaissance of another wonderfully harmonious period in the relationship between man and his environment, reminiscent of the heavenly countryside of southern England not so long ago. To be sure, sociologists and readers of Hardy will recall the unhappy lot of the farm worker and the animals, and the cruelties to which they were both heedlessly subjected; but this book is not primarily about people and livestock and pets; it is about the biosphere and the magic of Mother Earth. Enough of this pastoral landscape survives in Wessex to prove that some sort of harmony is still possible and to encourage the hope that it might even be extended as technology advances. As for the lot of the countryman, he has traded the cruel tyrannies of the past for the dubious comforts of a higher living standard at home and the noisy, smelly boredom of mechanized farming.

What, then, are those activities of man which pose a threat to the Earth and the life upon it? We as a species, aided by the industries at our command, have now significantly altered some of the major chemical cycles of the planet. We have increased the carbon cycle by 20 per cent, the nitrogen cycle by 50 per cent, and the sulphur cycle by over 100 per cent. As our numbers and our use of fossil fuels increase, these perturbations will grow likewise. What are the most likely consequences? The only thing we know to have happened so far is an increase in atmospheric carbon dioxide of about 10 per cent and perhaps also, although this is debatable, an increase in the burden of haze attributable to particles of sulphate compounds and soil dust.

It has been predicted that the increase in carbon dioxide will act as a sort of gaseous blanket to keep the Earth warmer. It has also been argued that the increase in haziness of the atmosphere might produce some cooling effect. It has even been suggested that at present these two effects cancel one another out and that is why nothing significant has so far emerged from the perturbation caused by the burning of fossil fuels. If the growth projections are correct and if as time goes by our consumption of these fuels continues to double more or less with each passing decade, we shall need to be viligant.

The parts of the Earth responsible for planetary control may still be those which carry the vast hordes of micro-organisms. The algae of the sea and of the soil surface use sunlight to perform the prime task of living chemistry, photosynthesis. They still turn over half of the Earth's supply of carbon, in co-operation with the aerobic decomposers of the soil and the sea-bed, together with the anaerobic microflora in the great mud zones of the continental shelves, sea bottom, marshes, and wet lands. The large animals, plants, and seaweeds may have important specialist functions, but the greater part of Gaia's self-regulating activity could still be conducted by micro-organisms.

As we shall see in the next chapter, there may be some regions of the world more vital to Gaia than others; so that however urgent the need to match the world's increasing population with increasing food supplies, we should take special care not to disturb too drastically those regions where planetary control may be sited. The continental shelves and wetlands generally have features and properties which make them suitable candidates for this role. It may be that we can create deserts and dust bowls with comparative impunity but if we devastate the area of the continental shelves through irresponsible bad husbandry in our first attempts at sea-farming, we shall do so at our peril.

Among the relatively few firm predictions about man's future is the one that our present numbers will, within the next few decades, at least double. The problem of feeding a world population of 8,000 million without seriously damaging Gaia would seem more urgent than that of industrial pollution. It may be argued, yes, but what of the more subtle poisons? The pesticides and herbicides, to say nothing of the ozone depleters, are surely the greatest threat?

A great debt is owed to Rachel Carson for having so movingly warned us of the threat to the biosphere arising from the careless and over-lavish use of pesticides. Yet it tends to be forgotten that we do take heed. A silent spring with no bird song has not arrived, although many birds, especially the rarer birds of prey, came near to extinction in some parts of the world. George Woodwell's careful study of the distribution and fate of DDT throughout the Earth is a model of how the pharmacology and toxicology of Gaia should be handled. The accumulation of DDT was not as great as was expected and recovery from its toxic effects was quicker. There appear to be natural processes for its removal which were not anticipated when the investigation began. The period of peak concentration of DDT in the biosphere is now well past. DDT will no doubt continue to be used in its life-saving and life-enriching role as a weapon against insect-borne disease, but it will probably be more carefully and economically employed in future. Such substances are like drugs, beneficial in the proper dose but harmful or even lethal in excess. It used to be said of fire, the first of the technological weapons, that it was a good servant but a bad master. The same holds true for the newer weapons of technology.

We may well need the fierce emotional drive of the radical environmentalist to alert us to the dangers of real or potential pollution hazards, but in our response we must take care not to over-react. In support of the campaign in the USA to ban all aerosol sprays, headlines such as 'The "Death Spray" that menaces every American' have appeared, followed by the warning, 'Those "harmless" spray cans may destroy all life on earth'. This kind of wild exaggeration may be good politics, but it is bad science. We must not throw out the baby with the bath-water—indeed, as the environmentalists would hasten to tell us, it is no longer acceptable to throw out even the bath-water; it must be recycled.

What of the fashionable contemporary doom by pollution, the erosion of the Earth's fragile shield against the deadly ultra-violet radiation of the sun? We are indebted to Paul Crutzen and Sherry Rowland for having alerted us to the potential threat to the ozone layer arising from the nitrogen oxides and the chlorofluorocarbons.

At the time of writing, ozone in the stratosphere continues its wavering but obstinate increase in density, as if unaware that it is supposed to be decreasing. Yet the arguments presented for its eventual depletion by pollutants are so convincing and reasonable that both law-makers and atmospheric scientists are concerned and uncertain as to the best course of action. Here Gaian experience may point the way. If the calculations by aeronomists are correct, then many natural events in the past should have profoundly depleted the ozone layer. For example, a major volcanic eruption such as Krakatoa in 1895 is likely to have injected vast quantities of chlorine compounds into the stratosphere which it is estimated could have depleted ozone by as much as 30 per cent. This figure represents at least twice the extent of depletion which is expected to take place by the year 2010 if chlorofluorocarbons continue to be released into the air at their present rate. Other untoward events include solar flares, large meteoric collisions, magnetic field reversals of the Earth, the supernova explosions of nearby stars, and possibly even the pathological over-production of nitrous oxide in the soil and the sea. Some or all of these incidents must have occurred with relative frequency in the past and will have generated in the stratosphere large volumes of the nitrogen oxides which are claimed to destroy the ozone. The survival of our own species and of the rich variety of life throughout Gaia seems conclusive evidence either that ozone depletion cannot be as lethal as it is often made out to be or that the theories are wrong and it never was depleted. Moreover, during the first 2,000 million years of life's presence on Earth there was no ozone at all and surface life, the bacteria and blue-green algae, must have been exposed to the full unshielded flood of ultra-violet from the sun.

We must not ignore those who would warn us, with their fearful tales of loathsome cancers resulting from the continued use of aerosol sprays and other contrivances, such as refrigerators, which contain chlorofluorocarbons. We must also not be panicked (as were the Federal Agencies of the USA) into premature and unjustified legislation banning the use of otherwise valuable and harmless products. Even on the gloomiest prediction, ozone depletion is a slow process. There is therefore ample time and every inclination on the part of scientists to investigate and prove or disprove the allegations, and then leave it to the law-makers to decide rationally what should be done.

We might also stop to think why it is that large quantities of nitrous oxide and of methyl chloride enter the atmosphere from biological sources, since both these compounds are claimed to be potent ozone depleters. At present it is thought that the ozone layer is at least 15 per cent thinner than it would be if these compounds of biological origin were not present in our atmosphere. As I have suggested in an earlier chapter, it may be that too much ozone is as harmful as too little and the production of nitrous oxide or methyl chloride from natural sources represents part of a Gaian regulatory system.

We are now well aware of the possible dangers of global pollution both of the atmosphere and of the oceans. National and international agencies are in the course of establishing monitoring stations equipped with sensors which will keep a record of the health of our planet. Satellites circling the Earth carry instruments to monitor the atmosphere, the oceans, and the land surface. So long as we can maintain a fairly high level of technology, this sensing programme is likely to continue and may even be extended. If the technology fails, then presumably other sectors of industry will also have failed and the potentially injurious effects of industrial pollution will diminish accordingly. In the end we may achieve a sensible and economic technology and be more in harmony with the rest of Gaia. I think that we are more likely to achieve this goal by retaining but modifying technology than by a reactionary 'back to nature' campaign. A high level of technology is by no means always energy-dependent. Witness the bicycle, the hang glider, a modern sailing craft, or a mini-computer performing in minutes man-years of calculation yet using less electricity than does a light bulb.

Our uncertainties about the future of our planet and the consequences of pollution stem largely from our ignorance of planetary control systems. If Gaia does indeed exist, then there are associations of species which co-operate to perform some essential regulatory functions. The thyroid gland is present in all mammals and most vertebrates. It harvests the meagre supplies of iodine from the internal bodily environment and converts them into iodine-bearing hormones which regulate our metabolism and without which we cannot live. As indicated in chapter 6, certain large marine algae, laminaria, may perform a similar function to the thyroid gland but on a planetary scale. These long straps of seaweed, whose habitat is the inshore waters where the sea always covers them even at the lowest tide, concentrate the element iodine from sea-water and process from their harvest a curious set of iodine-bearing substances. Several of these iodine compounds are volatile and escape into the sea and thence into the atmosphere. Prominent among them is methyl iodide. This substance when pure is a volatile liquid which boils at 42°C. It is very poisonous and almost certainly mutagenic and carcinogenic. Strange to relate, if it were an industrial product its presence might prohibit sea-bathing, according to United States legislation. The concentration of methyl iodide in the inshore waters and the air above is easily measurable with the extremely sensitive apparatus we now possess, and the United States law says that there must be no exposure to material which contains detectable amounts of a known carcinogen. Fear not! The present levels of methyl iodide in and around the sea are certainly and obviously tolerated by the creatures of that environment. Sea birds, fish, and seals may suffer from many things but not from the effects of locally produced methyl iodide. Nor is it likely that our occasional bathing in the sea will bring harm to us from this direction.

Methyl iodide produced by laminaria either escapes eventually into the atmosphere or reacts with sea-water to form a more chemically stable and even more volatile substance, methyl chloride. The methyl iodide which escapes from the sea travels through the air, but in a matter of hours, especially in sunlight, it is decomposed and sets free that life-essential element, iodine. Fortunately, iodine is also a volatile substance and remains in the air long enough to be blown across the continents. Some of it is believed to react with organic compounds in the air and re-form methyl iodide, but one way or another the iodine of the sea, concentrated by laminaria, is blown through the air to the land surfaces of the Earth and is absorbed by mammals like ourselves, who cannot live healthily without it. The algae which perform this vital function exist along a thin line surrounding the continents and islands of the world. The open ocean is by comparison a desert in which sea life is sparse indeed. It is important in a Gaian sense to think of the open oceans as a kind of marine Sahara and to keep in mind that the abundant life of the sea is concentrated in the inshore waters and above the continental shelves.

When I hear that there are proposals for the large-scale farming of kelp, which is a common name for laminaria, I find the prospect more disturbing than the possible effects of any of the industrial hazards which we have discussed. Kelp is the source of many useful products, apart from iodine. Alginates, for example, those sticky natural polymers, are valuable additives in a variety of industrial and domestic goods. If inshore farming is actively practised on the same scale as the land is now cultivated, uncomfortable consequences might follow both for Gaia and for us as an integral species.

A vast increase in kelp production might increase the flux of methyl chloride (the natural equivalent of the aerosol-propellant gases) and create a problem almost identical to that alleged to be the consequence of the release of the fluoro-chlorocarbons.

The breeding of strains of kelp which gave better yields of alginate would be an early step in farming practice. Such strains might lose their capacity to harvest iodine from the sea, or conversely their production of methyl iodide as well as their yield of alginate might reach levels toxic to other forms of inshore life.

Then there is the normal tendency of farmers to prefer monocultures. The kelp farmer would probably regard other algae as weeds and the herbivores of the inshore zone as vermin or pests threatening his profits. He would do his best, and his best is often very impressive, to destroy them. This sort of elimination may not matter so much on the land surfaces of the Earth which are the recipients of the sea's bounty; but this bounty is mainly produced within the continental shelves and inshore waters by a variety of species performing other essential services similar in kind but clearly different from the function of laminaria. The alga *Polysiphonia fastigiata* extracts sulphur from the sea and converts it to dimethyl sulphide, which subsequently reaches the atmosphere and is probably the normal natural carrier of sulphur in the air. An as yet unidentified species performs a similar task with selenium, another essential trace element for land mammals. It might be disastrous if these 'weeds' of the sea were eliminated in the interests of more intensive kelp farming.

The continental shelves cover a vast area, at least that of the continent of Africa. As yet, the farming of these regions is on a negligible scale, but we must not forget how rapidly mineral exploration has led to the successful establishment of oil and gas extraction plants for mining the fuel fields beneath the continental shelves. Once a resource is recognized, it does not take our species long to exploit it to the full.

As we saw in chapter 5, the continental shelves may also be vital in the regulation of the oxygen–carbon cycle. It is through the burial of carbon in the anaerobic muds of the sea-bed that a net increment of oxygen in the atmosphere is ensured. Without carbon burial, which leaves one additional oxygen molecule in the air for each carbon atom thus removed from the cycle of photosynthesis and respiration, oxygen would inevitably decline in concentration in the atmosphere until it almost reached vanishing-point. This danger is of no conceivable contemporary significance; indeed, it would take tens of thousands of years, or even more, to diminish oxygen in the atmosphere to any appreciable extent. Nevertheless oxygen regulation is a key Gaian process and the fact that it occurs on the continental shelves of the Earth emphasizes their singular importance. Knowing or perhaps even suspecting as much as we do now, it seems unwise to tamper with these regions. In view of what we still do not know about them, it may even be perilous.

The 'core' regions of Gaia, those between latitudes 45° North and 45° South, include the tropical forests and scrub lands. We may also need to keep a close eye on these areas if we are to guard against unpleasant surprises. It is well recognized that the agriculture of the tropical belt is often inefficient and that large stretches are already worked out or are being devastated through the same sort of primitive farming methods which led to the Bad Lands of the United States. What is less well known is that this bad farming is also disturbing the atmosphere on a global scale and to an extent at least comparable with the effects of urban industrial activity.

It is a common practice to clear scrub and forest land by burning, and also to burn off grass each year. Fires of this type inject into the air, in addition to carbon dioxide, a vast range of organic chemicals and a huge burden of aerosol particles. It is probable that the bulk of the chlorine now in the atmosphere is in the form of the gas methyl chloride, a direct product of tropical agriculture. Grass and forest fires generate at least five million tons of this gas each year, a far greater amount than is released by industry and probably also greater than the natural influx from the sea.

Methyl chloride is but one substance which we now know to be produced in abnormal quantities as a consequence of primitive agriculture. The brutal disturbance of natural ecosystems always involves the danger of upsetting the normal balance of atmospheric gases. Changes in the production rate of gases such as carbon dioxide or methane and of aerosol particles may all cause perturbations on a global scale. Moreover, even if Gaia is there to regulate and modify the consequences of our disruptive behaviour, we should remember that the devastation of the tropical ecosystems might diminish her capacity to do so.

It seems therefore that the principal dangers to our planet arising from man's activities may not be the special and singular evils of his urbanized industrial existence. When urban industrial man does something ecologically bad he notices it and tends to put things right again. The really critical areas which need careful watching are more likely to be the tropics and the seas close to the continental shores. It is in these regions, where few do watch, that harmful practices may be pursued to the point of no-return before their dangers are recognized; and so it is from these regions that unpleasant surprises are most likely to emerge. Here man may sap the vitality of Gaia by reducing productivity and by deleting key species in her life-support system; and he may then exacerbate the situation by releasing into the air or the sea abnormal quantities of compounds which are potentially dangerous on a global scale.

The European, American, and Chinese experience suggests that, given wise husbandry, twice the present human population of the world could be supported without uprooting other species, our partners in Gaia, from their natural habitats. It would be a grave mistake, however, to think that this could be achieved without a high degree of technology, intelligently organized and applied.

In the long run we have to guard against the dismal possibility that Rachel Carson was right but for the wrong reason. There may well come a silent spring bereft of bird song with the birds victims of DDT and other pesticides. If this does happen it will not be a consequence of their direct poisoning by the pesticides but because the saving of human lives by these agents will have left no room, no habitat, on Earth for the birds. As Garrett Hardin has said, the optimum number of people is not as large as the maximum the Earth can support; or, as it has been more bluntly expressed, 'There is only one pollution . . . People.'

CHAPTER 8

Living within Gaia

Some of you may have wondered how it has been possible to travel so far in a book about relationships amongst living things with only a brief mention of ecology. In the *Concise Oxford Dictionary* ecology is defined as: 'Branch of biology dealing with organisms' relations to one another and to their surroundings; (human) ecology, study of interaction of persons with their environment.' One purpose of this chapter is to consider Gaia in the light of human ecology, but first a brief word about recent developments in this subject.

Among the many distinguished contemporary human ecologists, there are two who represent most clearly the alternative policies which might guide mankind in its dealings with the rest of the biosphere. René Dubos has powerfully expressed the concept of man as the steward to life on Earth, in symbiosis with it like some grand gardener for all the world. It is a hopeful, optimistic view and a liberal one. In contrast to Dubos, Garrett Hardin apparently sees man as acting out a great tragedy which may lead not only to his own destruction but to that of the whole world. He suggests that our only means of escape is to renounce most of our technology, especially nuclear energy, but he seems to doubt whether we have free choice.

These two viewpoints encompass most of the current debating ground among human ecologists about the condition of mankind. To be sure, there are the many small fringe groups, mostly anarchist in flavour, who would hasten our doom by dismantling and destroying all technology. It is not clear whether their motivation is primarily misanthropic or Luddite, but either way they seem more concerned with destructive action than with constructive thought.

It can perhaps now be seen why we have not previously

discussed Gaia within the context of any branch of ecology. Whatever this science may have been originally, it has grown ever more associated in the public mind with human ecology. The Gaia hypothesis, on the other hand, started with observations of the Earth's atmosphere and other inorganic properties. Where life is concerned, it focuses special attention on what most people consider to be the lowest part, that represented by the micro-organisms. The human species is of course a key development for Gaia, but we have appeared so late in her life that it hardly seemed appropriate to start our quest by discussing our own relationships within her. Contemporary ecology may be deeply embedded in human affairs, but this book is about the whole of life on Earth within the older and more general framework of geology. Still, the nettle, a most unecological vegetable, bristling with poison barbs, must now be grasped. How then should we live within Gaia? What difference does her presence make to our own relationships with the world and with one another?

I suggest we start by considering in more detail Garrett Hardin's philosophy. In fairness to him it should be emphasized that his form of pessimism does not necessarily imply fatalism. It is rather, to use his own newly minted word, one of 'pejorism'. This means a stoical acceptance of the apocryphal Murphy's law: 'If anything can go wrong, it will', and implies a programme for the future based on the realistic awareness of this law and of the fact that we live in a very unfair universe. Perhaps the key to Hardin's view of life, and also to a good deal of present-day ecological thinking, is expressed in his quotation of the paraphrase of the three laws of thermodynamics:

'We can't win.'
'We are sure to lose.'
'We can't get out of the game.'

This set of laws, according to Hardin, is worse than pejoristic, it is tragic, since the essence of tragedy is that there is no escape. From the laws of thermodynamics there can be no escape, for they rule the whole of our universe, and we know of no other.

In such a context it is all too easy to accept as inevitable the use of nuclear weapons and other lethal products of technology in the absurd and truly tragic battles between tribal groups. The battles are waged under cover of such high-sounding slogans as justice, liberation, or national self-determination, which serve only to cover the true motivations of greed, power, and envy. This kind of double talk being all too human and widespread, it is easy to understand the violence of the environmental movement's protest against proposed extensions of atomic power plants, and the ecologists' distrust of the claims that they are solely for peaceful purposes.

Much of this book was written in Ireland, where tribal warfare has never been far from the scene. Yet paradoxically Hardin's forebodings have much less reality in the relaxed and informal atmosphere of Irish country life than they do in structured, highly organized urban societies. As the saying is, the further away from the battle, the fiercer the patriotic fervour.

Let us again look at the laws of thermodynamics. It is true that at first sight they read like the notice at the gate of Dante's Hell; but in fact, tough as they are and although like income tax they cannot without penalty be evaded, they can with forethought be avoided. The Second Law states unequivocally that the entropy of a closed system must increase. Since we are all closed systems, this means that all of us are doomed to die. Yet it is so often ignored or deliberately forgotten that the unending death-roll of all creatures, including ourselves, is the essential complement to the unceasing renewal of life. The death sentence of the Second Law applies only to identities, to closed systems, and could be rephrased: 'Mortality is the price of identity.' The family lives longer than its members, the tribe longer still, and *homo sapiens* as a species has existed for several million years. Gaia, the sum of the biota and those parts of the environment coming under its influence, is probably three and a half aeons old. This is a most remarkable yet quite legal avoidance of the Second Law. In the end, the sun will overheat and all life on Earth will cease, but that may not happen before several more aeons have passed. Compared with the lifetime of our species, let alone that of an individual human being, this time span is no tragic brief spell, but offers almost an infinity of opportunities to terrestrial life. Whoever set up the rules of this universe clearly had no time for those who cry unfair. The prizes, though, are only to be won by the cunning, bold, and resolute, with the wit to seize on any chance that offers.

It is pointless to blame the universe and its laws for defects in the human condition. If it offends the moral sense to be born into some cosmic Las Vegas with unbreakable house rules and with no chance of escape, think instead how wonderful it is that we have survived as a species in a world of one-arm bandits and still have a chance to take stock and plan our future tactics. Consider the odds against this, three and a half aeons ago. They would have raised doubts even in the mind of that super-optimist, Dr Pangloss, who believed so firmly that we were born into the best of all possible worlds. To be sure, the Second Law says that you never can win, that you are bound to die, but in the fine print it also says that almost anything can happen while you stay in the game. Although one can be deeply moved, as I am, by Hardin's thoughts and words and by the hauntingly beautiful imagery he often uses, one must not ignore the fact that he is concerned with human ecology rather than with all of the biosphere.

In science, simultaneous macroscopic and microscopic exploration is quite customary, especially in biology. Molecular biology, for example, which derived from the application of chemical analysis to biological problems and led to the discovery of DNA and its function as the carrier of genetic information for every form of life, has developed independently from physiology, which concerns the whole animal and the way it functions as an integrated living system. In like manner, the difference between the Gaian notion and the ecological notion of our planet derives in part from their history. The start of the Gaia hypothesis was the view of the Earth from space, revealing the planet as a whole but not in detail. Ecology is rooted in down-to-Earth natural history and the detailed study of habitats and ecosystems without taking in the whole picture. The one cannot see the trees in the wood. The other cannot see the wood for the trees.

If we assume Gaia's existence, we can make other assumptions which shed a new light on our place in the world. For example, in a Gaian world our species with its technology is simply an inevitable part of the natural scene. Yet our relationship with our technology releases ever-increasing amounts of energy and provides us with a similarly increasing capacity to channel and process information. Cybernetics tells us that we might safely pass through these turbulent times if our skills in handling information develop faster than our capacity to produce more energy. In other words, if we can always control the genie we have let out of the bottle.

An increase of power input to a system may enhance the loop gain and so assist in the maintenance of stability, but if the response is too slow, increasing the power input could be the recipe for a whole series of cybernetic disasters. Imagine a world with the present-day arsenal of nuclear weapons but with no means whatever of telecommunication. A key factor in our relationships with the rest of the world and with each other is our capacity to make the correct response in time.

Having assumed her existence, let us consider three of Gaia's principal characteristics which could profoundly modify our interaction with the rest of the biosphere.

1. The most important property of Gaia is the tendency to optimize conditions for all terrestrial life. Provided that we have not seriously interfered with her optimizing capacity, this tendency should be as predominant now as it was before man's arrival on the scene.
2. Gaia has vital organs at the core, as well as expendable or redundant ones mainly on the periphery. What we do to our planet may depend greatly on where we do it.
3. Gaian responses to changes for the worse must obey the rules of cybernetics, where the time constant and the loop gain are important factors. Thus the regulation of oxygen has a time constant measured in thousands of years. Such slow processes give the least warning of undesirable trends. By the time it is realized that all is not well and action is taken, inertial drag will bring things to a worse state before an equally slow improvement can set in.

For the first of these characteristics, we have assumed that the Gaian world evolves through Darwinian natural selection, its goal being the maintenance of conditions optimal for life in all circumstances, including variations in output from the sun and from the planet's own interior. We have in addition made the assumption that from its origin the human species has been as much a part of Gaia as have all other species and that like them it has acted unconsciously in the process of planetary homoeostasis.

However, in the past few hundred years our species, together with its dependent crops and livestock, has grown in numbers to occupy a substantial proportion of the total biomass. At the same time the proportion of energy, information, and raw materials which we use has grown at an even faster rate through the magnifying effect of technology. It therefore seems important in the context of Gaia to ask: 'What has been the effect of all or any of these recent developments? Is technological man still a part of Gaia or are we in some or in many ways alienated from her?'

I am grateful to my colleague Lynn Margulis for demystifying these most difficult questions about Gaia by observing that: 'Each species to a greater or lesser degree modifies its environment to optimize its reproduction rate. Gaia follows from this by being the sum total of all of these individual modifications and by the fact that all species are connected, for the production of gases, food and waste removal, however circuitously, to all others.' In other words, like it or not, and whatever we may do to the total system, we shall continue to be drawn, albeit unawares, into the Gaian process of regulation. Since we are not yet a completely social species, we may be participating on both the community and personal levels.

If it should seem far-fetched to expect that any actions by individuals or by the community could halt the growing depredation of the Earth or influence such serious problems as population growth, consider what has happened in the past twenty years, since we became aware of ecological problems global in scale. During this short period new laws and regulations have been introduced in most countries, limiting the freedom of entrepreneurs and of industry in the interest of ecology and the environment. Indeed, the negative feedback has been great enough seriously to affect economic growth.

Very few, if any, of the pundits and soothsayers of the early nineteen-sixties predicted that by now the environmental movement would be exerting a curb on economic expansion. Yet it does, and not only through direct measures, as for example requiring industry to channel some of its profits into expenditure on clearing up the waste it produces. A further loss of growth potential arises from the need to divert research and development away from the introduction of new products into efforts to solve environmental problems. An ecological cause may not always be as valid as, for instance, the protest over the abuse of pesticides, turning useful and efficient means of pest control into indiscriminate weapons against the biosphere. Some ecologists warned that there were flaws in the original design of the Alaskan pipeline which was to convey oil to the United States. Their objections were sensible but soon an extraordinary and hypocritical agonizing by others so effectively delayed its construction that the energy shortages of 1974 were largely the result of this, and not, as is usually claimed, a consequence of the price increase imposed by the other oil-producing nations. The cost of this delay is estimated as being $30,000 million. The exploitation of human ecology for political ends can become nihilistic, rather than a force working for reconciliation between mankind and the natural world.

Coming to our second characteristic, what regions of the Earth are vital to Gaia's well-being? Which ones could she do without? On this subject we already have some useful information. We know that the regions of the globe outside latitudes 45° North and 45° South are subject to glaciations, when vast overburdens of ice and snow all but sterilize the land and in places bulldoze the soil away down to the bedrock itself. Even though most of our industrial centres are in the northern temperate regions which are subject to glaciations, nothing we have done so far by way of industrial scarring and pollution in these areas can equal the natural devastations of the ice. It seems, therefore, that Gaia can tolerate the loss of these parts of her territory, about 30 per cent of the Earth's surface, although her present losses are somewhat less, since in between glaciations there are still regions of ice and permafrost.

However, in times past the fertile regions of the tropics were unaffected by man and could therefore have made good the losses suffered during ice ages. Can we be sure that another Ice Age could be endured, once the core regions of the Earth's surface have been denuded of their forests, as they may well be in a few decades from now? It is all too easy to think that environmental and pollution problems are to be found solely among the industrial nations. It was timely to have no less an authority than Bert Bolin list the extent and rapidity of the destruction of the tropical forests, and also discuss some of the possible consequences of their loss. Even though man survives, we can be certain that the total destruction of the intricate and contrived tropical forest ecosystems is a loss of opportunities for all creatures on Earth.

It will no doubt be decided in due course by natural selection which is the most fit to survive: a maximum population of humans living at bare subsistence level in a semi-desert—the ultimate welfare world—or some other less costly social system with fewer people. It could be argued that a world with tens of thousands of millions of human beings on its surface is not only possible but tolerable, through the continued development of technology. The amount of regimentation, self-discipline, and sacrifice of personal freedom that would of necessity be imposed on everyone in such a crowded world must make it unacceptable to many by our present standards. We should bear in mind, however, that conditions in present-day China and Britain both indicate that high-density living is neither impossible nor always unpleasant. To succeed on a world scale, a clear understanding and knowledge of our territorial limits within Gaia would be essential, and the most scrupulous care would need to be taken to maintain the integrity of those key regions which are found to regulate the planetary health.

If we are lucky we may find that the vital organs in the body of Gaia are not on the land surfaces but in the estuaries, wet lands, and muds on the continental shelves. There, the rate of carbon burial adjusts automatically to regulate the concentration of oxygen, and essential elements are returned to the atmosphere. Until we know much more about the Earth and the role of these regions, vital or otherwise, we had much better set them outside the limits for exploitation.

There may, of course, be other and unexpected vital areas. We do not know, for instance, how important is the output of methane to the air from anaerobic micro-organisms. As indicated in chapter 5, methane production may be significant in the control of oxygen, but some of these anaerobic communities live not on the sea bed but in our guts and in those of other animals. In his pioneering research into the biochemistry of the atmosphere, Hutchinson suggested that nearly all atmospheric methane might come from this source. It could just be true that at some time the additional amount of methane and other gases produced in our guts made all the difference. A flight of farts, you may think, but it serves to illustrate how little we know about this whole subject. It also reminds us that we may at times fulfil quite lowly functions in Gaia's living system, whatever notions we may entertain about our ultimate preferment.

A detailed examination of the cycles which regulate the atmospheric concentration of oxygen illustrated in chapter 5 will reveal a network of intricate loops, one so complex as still to defy complete analysis. This brings us to the third property of Gaia, namely that she is a cybernetic system. The many pathways involved in regulation can have associated with them different time constants and different functional capacities, or as the engineer might say, variable loop gains. The larger the proportion of the Earth's biomass occupied by mankind and the animals and crops required to nourish us, the more involved we become in the transfer of solar and other energy throughout the entire system. As the transfer of power to our species proceeds, our responsibility for maintaining planetary homoeostasis grows with it, whether we are conscious of the fact or not. Each time we significantly alter part of some natural process of regulation or introduce some new source of energy or information, we are increasing the probability that one of these changes will weaken the stability of the entire system, by cutting down the variety of response.

In any operating system whose goal is homoeostasis, departures from the current optimum caused by changes in the energy fluxes or their response times will tend to be corrected and a new optimum sought which incorporates the changes. A system as experienced as Gaia is unlikely to be easily disturbed. Nevertheless, we shall have to tread carefully to avoid the cybernetic disasters of runaway positive feedback or of sustained oscillation. If, for example, the methods of climate control which I have postulated were subject to severe perturbation, we might suffer either a planetary fever or the chill of an ice age, or even experience sustained oscillations between these two uncomfortable states.

This could happen if, at some intolerable population density, man had encroached upon Gaia's functional power to such an extent that he disabled her. He would wake up one day to find that he had the permanent lifelong job of planetary maintenance engineer. Gaia would have retreated into the muds, and the ceaseless intricate task of keeping all of the global cycles in balance would be ours. Then at last we should be riding that strange contraption, the 'spaceship Earth', and whatever tamed and domesticated biosphere remained would indeed be our 'life support system'. No one yet knows what is the optimum number for the human species. The analytic equipment needed to provide the answer is not yet assembled. Assuming the present per capita use of energy, we can guess that at less than 10,000 million we should still be in a Gaian world. But somewhere beyond this figure, especially if the consumption of energy increases, lies the final choice of permanent enslavement on the prison hulk of the spaceship Earth, or gigadeath to enable the survivors to restore a Gaian world.

What is remarkable about man is not the size of his brain, no greater than that of a dolphin, nor his loose incomplete development as a social animal, nor even the faculty of speech or his ability to use tools. Man is remarkable because by the combination of all of these things he has created an entirely new entity. When socially organized and equipped with a technology even as rudimentary as that of a Stone Age tribal group, man has the novel capacity to collect, store, and process information, and then use it to manipulate the environment in a purposeful and anticipatory fashion.

When primates followed the evolutionary steps of the ant and first formed an intelligent nest, its potential for changing the very face of the Earth was as revolutionary as that of photosynthetic oxygen producers when they first emerged aeons before. From the very beginning this new organization had the capacity to modify the environment on a global scale. There is good evidence, for example, to show that the arrival of infant mankind in North America after the crossing of the Bering Strait led to the elimination of a range of animal species, mostly large mammals, on a continental scale and in only a very few years. This was the era of the cruel and lazy technique of fire-drive hunting, when food for the tribe was gathered by setting fire to the forest along a line front, moving to a convenient site down-wind and then waiting for the fire to drive the food straight into the sticks and spears of the hunters. By any reckoning this was at that time an ecologically disastrous application of new technology, and yet, as Eugene Odum has reminded us, its application led to the development and evolution of the great grassland ecosystems.

If we scan backwards over the history of man as a collective species and direct our attention particularly to his relationships with the global environment, we discern a series of repetitions. There are periods of rapid technological development leading to what seems to be an environmental catastrophe. This is followed by a quite lengthy period of stability and coexistence with a new and modified ecosystem. Fire-drive hunting, as we have seen, led to the destruction of forest ecosystems, but was followed by the establishment of the great grassland ecosystems, the savannahs, and a new period of coexistence. For a more recent example, Dubos reminds us that the Acts of Enclosure in England, which denied access to the common land and led to the characteristically English landscape with its rich hedgerow habitats, were regarded at the time of their introduction as an environmental disaster. The destruction of the hedgerows when farming evolved to become 'agribusiness' is now much mourned, but again, as Dubos properly asks, will not the new ecosystem which comes to terms with the agribusinessmen be mourned in its turn when it gives way to some fresh technological advance? This pro-

gression might be defined as 'grandfather's law' which states
that 'things were better in the old days'.

It is a fact of life that new evolutionary developments cause
distress to the established order. This is so at all levels of life.
There is the example at the lowest level of the mutation of a
virus from one which causes discomfort to one that is lethal.
This happened with a strain of influenza virus in 1918, when
the 'Spanish Flu' pandemic caused more deaths than the total
number killed in the First World War. There is also the
successful new organization of the fire ant which enabled it to
invade and colonize the North American continent. Anyone
who has had the misfortune to disturb a fire ant's nest must
know how agonizingly unwelcome this invader is.

Our continuing development as an intelligent social animal
with an ever-increasing dependence upon technology has
inevitably disturbed the rest of the biosphere and will continue
to do so. The mutation rate of man himself may be very slow
but the rate of change of the collective association which
constitutes mankind is increasing all the time. Richard Daw-
kins has observed that both major and minor technological
advances can be regarded as analogous to mutations in this
context.

The remarkable success of our species derives from its
capacity to collect, compare, and establish the answers to
environmental questions, thus accumulating what is sometimes called conventional or tribal wisdom. Originally handed
down by word of mouth from generation to generation, this
wisdom has now become a bewildering mass of stored information. In a small tribal group still living in its natural habitat,
the interaction with the environment is intense and where
conventional wisdom and Gaian optimization conflict, the
discrepancy is rapidly seen and the correction made. This may
be why such groups as the Eskimos and the Bushmen appear to
lead well-adapted lives, optimal for their extreme and unusual
environments. It is a commonplace that infection by the larger,
more diffuse wisdom of our own urban and industrial societies
has in general been harmful to them. Many of us have seen
those sad and moving films of 'civilized' Eskimos sitting in
prefabricated huts, chain-smoking cigarettes and bemoaning

the fate of their children, who have been taken from them to be taught the three Rs instead of how to live in the Arctic.

As society became more urbanized, the proportion of information flow from the biosphere to the pool of knowledge which constitutes the wisdom of the city decreased, compared with the proportion entering the wisdom of rural or hunting communities. At the same time the complex interactions within the city produced new problems requiring attention. These were resolved and their solutions stored. Soon city wisdom became almost entirely centred on the problems of human relationships, in contrast to the wisdom of any natural tribal group, where relationships with the rest of the animate and inanimate world are still given due place.

I have often wondered about the allegation that Descartes likened animals to machines because they had no souls, whereas man with his immortal soul was sentient and capable of rational thought. Descartes was a highly intelligent man, and it seems incredible that he could have been so unobservant as to believe that pain could only be consciously felt by man and that cruelty to a horse or a cat was of no consequence because they had no more realization of pain than an inanimate object such as a table. Whether or not this was his belief, the dreadful notion was credible to many in his time and has had a long run since. It illustrates the extent to which the conventional wisdom of a closed urban society becomes isolated from the natural world. Let us hope that this alienation is coming to an end and that the many splendid natural-history and wild-life films shown on television may help to dispel it. We are now in the midst of a communications explosion, and television will soon provide everyone with a window on the world. It has already enormously extended and increased the size, rate, and variety of the information flow. We may at last be moving away from the stagnant reaches of a society with its roots in medieval life.

So far in this chapter we have dealt with what might go wrong in the future rather than with what might go right, but there is a more optimistic point of view. Most newspapers carry life assurance advertisements which attract the young and early middle-aged with promises of quite substantial sums accruing to them in their sixties, all for some modest monthly premium. It is a tribute to the faith of most people in the future that insurance companies still do good business. That great prophet of the future, Herman Kahn, sees an America of 600 millions in the next century, mostly living at the population density of Westchester County, 2,000 per square mile. He believes, and convincingly argues, that all the essentials of life will be available indefinitely to support such a population and in a much more developed world than exists now. Indeed, almost all those professionals who gather information about the world's resources and study it with the aid of powerful analytical devices believe that the present trends of expansion in human population and technology will continue for at least another thirty years.

Most governments and many large multinational corporations now either buy the services of these forecasters or establish their own prediction units. These highly talented groups, equipped with some of the most powerful computers now available, gather and sift the information of the world and use the resulting data to build hypotheses, or models, as they are more usually called these days, which are then continually refined until the future can be seen with what appears to be no less clarity than were the pictures on our early television screens. Parallel with this new development in 'futurology', more and more scientists are conducting their researches by reference to similar models. Experimental measurements are made and entered in a computer, where they are compared with the predictions of a hypothesis. If there is disagreement, the evidence is examined for errors, or the model is discarded and another tried to find a better fit. When the scientist who gathers the experimental facts is also the model builder or a close colleague, this works very well indeed. The rapidity with which the computer can perform the otherwise brain-racking labour of endless arithmetic makes this a potent combination and the hypothesis most suited to promotion to a theory is soon selected. Unfortunately, most scientists live their lives in cities and have little or no contact with the natural world. Their models of the Earth are built in universities or institutions where there is all the talent and the hardware necessary, but what tends to be missing is that vital ingredient, information gathered first-hand in the real world. In these circumstances it is a natural temptation to assume that the information contained in scientific books and papers is adequate, and that if some of it does not fit the model then the facts must be wrong. From that point, the fatal step of selecting only data which fit the model is all too easy, and soon we have built an image not of a real world, which might be Gaia, but of that obsessive delusion, Galatea, Pygmalion's fair statue.

I speak from personal knowledge when I say that those of us who go forth in ships or travel to remote places to measure things about the atmosphere or the ocean and their interactions with the biosphere are few in number compared with those who choose to work in city-based institutions and universities. Personal contact between the explorers and the model builders is rare and information passes through the terse limited phraseology of scientific papers, where subtle, qualifying observations cannot be included along with the data. It is hardly surprising that the models are all too often Galatean.

If we are to live properly within Gaia, this imbalance needs early correction. A flow of accurate information about all aspects of the world is essential. Model building from yesterday's inadequate data is as absurd as forecasting tomorrow's weather using a giant computer and last month's input data. Adrian Tuck of the British Meteorological Office often reminds me how salutary it is to recall that the most experienced and professional of all prediction sciences is weather forecasting. Present-day weather forecasting makes use of the most comprehensive and reliable data-gathering network yet available, the most powerful computers in the world, and some of the most talented and able members of our society. Yet with how much certainty can the weather be predicted even a month ahead, let alone into the next century?

Just as a man who experiences sensory deprivation has been shown to suffer hallucinations, it may be that the model builders who live in cities are prone to make nightmares rather then realities. No one who has experienced the intense involvement of computer modelling would deny that the temptation exists to use any data input that will enable one to continue playing what is perhaps the ultimate game of solitaire.

As things are, our ignorance of the possible consequences of our actions is so great that useful predictions of the future are almost ruled out. The situation is made worse by the fact that the political polarization of our world and the break-up of society into small myopic tribal entities makes exploration and the gathering of scientific evidence increasingly difficult. None of the great journeys of exploration of the last century, such as the voyages of the *Beagle* or the *Challenger*, could now be accomplished without let or hindrance. Rightly or wrongly, research ships are often regarded by developing nations as agents of neo-colonial exploitation seeking mineral wealth on their continental shelves. In 1976 the Argentinians raised this type of resentment to a new level by firing on the *Shackleton* as she sailed near the Falkland Islands in the course of her scientific investigations. Similarly, it is now difficult for an independent observer to take equipment for atmospheric analysis into many tropical countries. Scientific investigation appears to have become nationalized; it must be done by a citizen of the country or not at all. Whether or not there is a real or historical justification for such fears of exploitation, they are undoubtedly widespread in that half of the world which lies in the tropics, and in consequence scientific investigation on a global scale is becoming increasingly difficult.

Although we may doubt whether the denizens of the think-tanks are making proper models of the world to come, one thing does seem certain about the near future: there can be no voluntary resignation from technology. We are so inextricably part of the technosphere that giving it up is as unrealistic as jumping off a ship in mid-Atlantic to swim the rest of the journey in glorious independence. There have been numerous groups who have tried to escape from modern society and go back to nature. Nearly all of them have failed, and when the rare partial successes are examined, strong support from the rest of us is always seen to be present. There is a Gaian analogy here, in that, as we have seen in chapter 6, where micro-organisms and sometimes more complex life forms have successfully colonized an extreme environment, such as a

spring of boiling water or a salt lake, they survive only because
the rest of us in Gaia maintain the supply of essentials such as
oxygen and nutrients. Just as the toleration of personal eccen-
tricity is the hallmark of a rich and successful civilization, so it
is with biological eccentricities. They can only happen on a
flourishing planet. (This is another reason, incidentally, why
the search for sparse life adapted to the harsh conditions of
Mars is probably in vain.) A more promising solution to the
problems we have created for ourselves is that of the alternative, or appropriate, technology movement. Here there is an
honest recognition of our dependence on technology and an
attempt to select only those parts of it which are seemly and
moderate in their demands on planetary resources.

In our attempts to resolve the crisis of diminishing
resources, we seem consistently to underestimate the facilities
of the press and of the telecommunications systems: not only
their capacity to influence events as a result of pressure on
other powerful institutions and groups, as in that formerly
much used phrase 'the power of the press', but also through
their technical capacity to inform the whole world about all
that is happening most of the time. As we have seen, the rapid
dissemination of information about the environment helps to
reduce the time constant of our response to adverse changes.

Not so long ago it seemed that mankind was like a cancer on
this planet. We had apparently severed the feedback loops of
pestilence and famine which regulated our numbers. We were
growing unrestrictedly at the expense of the rest of the
biosphere and at the same time our industrial pollution and
chemically contrived antibiotic agents like DDT were poisoning those few remaining creatures that we had not deprived of
their habitats. The danger is still there in places. Yet the
population no longer increases everywhere, industry is far
more conscious of its effect on the environment, and there is
above all growing public awareness of our situation. We might
claim that the spread of information about our problems is
leading to the development of new processes for controlling, if
not solving, them. It is all to the good that we no longer require
the brutal regulation of our numbers by disease and famine. In
many countries families are now limited on a purely voluntary

basis through the desire for a better quality of life, which is rarely possible at the maximum rate of child-bearing. We must never forget, of course, that this may be a temporary phase and that, as C. G. Darwin warns us, natural selection will ensure that under voluntary population control *Homo philoprogenitus* is bound to win, and when he does our numbers will grow again at an even faster rate.

The revolution in information technology is likely to change the future world in ways that none of us can now envisage. In a significant article in the *Scientific American* in 1970, Tribus and McIrvine developed the theme 'Knowledge is power' in a most comprehensive fashion. They showed among many other things that the beneficence of the sun could be regarded as a continuous gift of 10^{37} words of information per second to the Earth, rather than as 5×10^7 megawatt hours of power per second, as is the usual way of putting it. We have seen that we are near the limits of what can be done with this much energy, even solar energy, but there are almost no limits to our ability to harness this flood of information from our sun. With the aid of our newly invented hardware, we embark with growing delight on the first explorations of that rich world of information, idea space. Will this lead to yet another environmental disturbance? Has the pollution of idea space already begun in the haziness and increased entropy of language compared with what it used to be?

To everything there is a season, and a time to every purpose under the heaven; a time to be born, and a time to die; a time to plant, and a time to pluck up that which is planted.

I returned, and saw under the sun, that the race is not to the swift, nor the battle to the strong, neither yet bread to the wise, nor yet riches to men of understanding, nor yet favour to men of skill, but time and chance happeneth to them all.

Beauty is truth, truth beauty—that is all
Ye know on Earth, and all ye need to know.

There can be no prescription, no set of rules, for living within Gaia. For each of our different actions there are only consequences.

CHAPTER 9

Epilogue

My father was born in 1872 and raised on the Berkshire Downs just south of Wantage. He was an excellent and enthusiastic gardener and also a very gentle man. I remember him rescuing wasps from drowning after they had blundered into the water butt. He would say, 'They are there for a purpose, you know', and then explain to me how they controlled the aphids on his plum trees and how they were surely due some of the crops as a reward.

He had no formal religious beliefs and did not attend church or chapel. I think his moral system came from that unstructured mixture of Christianity and magic which is common enough among country people, and in which May Day as well as Easter Day is an occasion for ritual and rejoicing. He felt instinctively his kinship with all living things and I remember how greatly it distressed him to see a tree cut down. I owe much of my own feeling for natural things to walks with him down country lanes and along ancient drives which had, or appeared in those days to have, a sweet seemliness and tranquillity.

This chapter begins autobiographically so that I may bring us the more easily to consider the most speculative and intangible aspects of the Gaia hypothesis: those which concern thought and emotion in the interrelationship of man and Gaia.

Let us start by considering our sense of beauty. By this, I mean those complex feelings of pleasure, recognition, and fulfilment, of wonder, excitement, and yearning, which fill us when we see, feel, smell, or hear whatever heightens our self-awareness and at the same time deepens our perception of the true nature of things. It has often been said—and for some, *ad nauseam*—that these pleasurable sensations are inextricably bound up with that strange hyperaesthesia of romantic love. Even so, there seems no need inevitably to attribute the pleasure we feel on a country walk, as our gaze wanders over the downs, to our instinctive comparison of the smooth rounded hills with the contours of a woman's breasts. The thought may indeed occur to us, but we could also explain our pleasure in Gaian terms.

Part of our reward for fulfilling our biological role of creating a home and raising a family is the underlying sense of satisfaction. However hard and disappointing at times the task may have been, we are still pleasurably aware at a deeper level of having played our proper part and stayed in the mainstream of life. We are equally and painfully aware of a sense of failure and loss if for some reason or other we have missed our way or made a mess of things.

It may be that we are also programmed to recognize instinctively our optimal role in relation to other forms of life around us. When we act according to this instinct in our dealings with our partners in Gaia, we are rewarded by finding that what seems right also looks good and arouses those pleasurable feelings which comprise our sense of beauty. When this relationship with our environment is spoilt or mishandled, we suffer from a sense of emptiness and deprivation. Many of us know the shock of finding that some peaceful rural haunt of our youth where once the wild thyme blew and where the hedges were thick with eglantine and may, has become a featureless expanse of pure weed-free barley.

It does not seem inconsistent with the Darwinian forces of evolutionary selection for a sense of pleasure to reward us by encouraging us to achieve a balanced relationship between ourselves and other forms of life. The thousand-year-old New Forest in southern England, once the private hunting reserve of William the Conqueror and his Norman barons, is still an area of great scenic beauty, where badgers roam at night and ponies have right of way over humans and the internal combustion engine. Although this historic, 130-square-mile region of ancient woodland and heath is protected by special Acts of Parliament, the true price of its survival is our unceasing vigilance. For it is now the pleasure-ground of thousands of holiday picnickers, campers, and tourists, who drop 600 tons of litter annually and sometimes, with a careless match or cigarette, start fires which may destroy in a few hours over many acres the product of centuries-old balanced husbandry between the forester and his environment.

Another of our instincts which probably favours survival is that which associates fitness and due proportion with beauty in individuals. Our bodies are formed of cell co-operatives. Each nucleus-containing body cell is an association of lesser entities in symbiosis. If the product of all this co-operative effort, a human being, seems beautiful when correctly and expertly assembled, is it too much to suggest that we may recognize by the same instinct the beauty and fittingness of an environment created by an assembly of creatures, including man, and by other forms of life? Where every prospect pleases, and man, accepting his role as a partner in Gaia, need not be vile.

It would be dauntingly difficult to test experimentally the notion that the instinct to associate fitness with beauty favours survival, but it might be worth a try. I wonder if a positive answer would enable us to rate beauty objectively, rather than through the eye of the beholder. We have seen that the capacity greatly to reduce entropy or, to put it in the terms of information theory, greatly to reduce the uncertainty of the answers to the questions about life, is itself a measure of life. Let us set beauty as equal to such a measure of life. Then it could follow that beauty also is associated with lowered entropy, reduced uncertainty, and less vagueness. Perhaps we have always known this, since it is after all part of our internal life recognition programme. Because of it we, through the eye of Blake, even saw our predator as beautiful:

> Tiger! Tiger! burning bright
> In the forests of the night,
> What immortal hand or eye
> Could frame thy fearful symmetry?
>
> In what distant deeps or skies
> Burnt the fire of thine eyes?
> On what wings dare he aspire?
> What the hand dare seize the fire?

It might even be that the Platonic absolute of beauty does mean something and can be measured against that unattainable state of certainty about the nature of life itself.

My father never told me why he believed that everything in this world was there for a purpose, but his thoughts and feelings about the countryside must have been based on a mixture of instinct, observation, and tribal wisdom. These persist in diluted form in many of us today and are still strong enough to power environmental movements which have come to be accepted as forces to be reckoned with by other powerful pressure groups in our society. As a result, the churches of the monotheistic religions, and the recent heresies of humanism and Marxism, are faced with the unwelcome truth that some part of their old enemy, Wordsworth's Pagan, 'suckled in a creed outworn', is still alive within us.

In earlier times, when plague and famine regulated our numbers, it seemed fair and fitting to try by every means to heal the sick and preserve human life. This attitude later crystallized into the rigidly uncompromising belief that the Earth was made for man and his needs and desires were paramount. In authoritarian societies and institutions, it seemed absurd to doubt the wisdom or propriety of razing a forest, damming a river, or building an urban complex. If it was for the material good of human beings, then it must be right. No moral question was involved, other than the need to prevent bribery and corruption and to ensure fair shares among the beneficiaries.

The pangs that many people now feel at the sight of dunes, salt-marshes, woodlands, and even villages brutally destroyed and erased from the face of the Earth by bulldozers are very real. It is no comfort to be told that this attitude is reactionary and that the new urban development will provide jobs and opportunities for young people. The fact that this answer is partly true increases the sense of pain and outrage by denying a right to express it. In such circumstances it is hardly surprising that the environmental movement, although powerful, has no clear-cut objective. It tends to attack quite viciously such inappropriate targets as the fluorocarbon industry and fox-hunting, while turning a blind eye to the potentially more serious problems posed by most methods of agriculture.

The strong but confused emotions aroused by the worst excesses of public works and private enterprise provide ripe material for exploitation by unscrupulous manipulators. Environmental politics is a lush new pasture for demagogues and therefore an increasing source of anxiety to responsible governments and industries alike. Attaching that overworked adjective 'environmental' to the names of departments and agencies dealing with various aspects of the problem seems unlikely to stem the rising tide of anger and protest.

Biological arguments which appear to have a sound scientific basis are often used to support environmental causes, but usually they carry very little weight with scientists. Ecologists know that so far there is no evidence that any of man's activities have diminished the total productivity of the biosphere. Whatever an ecologist may feel as an individual about an imminent problem, his hands are tied by a lack of hard scientific evidence. The result is an environmental movement which is thwarted, bewildered, and angry.

The churches and the humanist movements have sensed the powerful emotional charge generated by the environmental campaign and have re-examined their tenets and beliefs so as to take account of it. There is, for example, a fresh awareness of the concept of Christian stewardship whereby man, while still allowed dominion over the fish and the fowl and every living thing, is accountable to God for the good management of the Earth.

From a Gaian viewpoint, all attempts to rationalize a subjugated biosphere with man in charge are as doomed to failure as the similar concept of benevolent colonialism. They all assume that man is the possessor of this planet; if not the owner, then the tenant. The allegory of Orwell's *Animal Farm* takes on a deeper significance when we realize that all human societies in one way or another regard the world as their farm. The Gaia hypothesis implies that the stable state of our planet includes man as a part of, or partner in, a very democratic entity.

Among several difficult concepts embodied in the Gaia hypothesis is that of intelligence. Like life itself, we can at present only categorize and cannot completely define it. Intelligence is a property of living systems and is concerned with the ability to answer questions correctly. We might add, especially questions about those responses to the environment which affect the system's survival, and the survival of the association of systems to which it belongs.

At the cellular level, decisions as to the edibility or otherwise of things encountered, and as to whether the environment is favourable or hazardous, are vital for survival. They are, however, automatic processes and do not involve conscious thought. Much of the routine operation of homoeostasis, whether it be for the cell, the animal, or for the entire biosphere, takes place automatically, and yet it must be recognized that some form of intelligence is required even within an automatic process, to interpret correctly information received about the environment. To supply the right answers to simple questions such as: 'Is it too hot?' or: 'Is there enough air to breathe?' requires intelligence. Even at the most rudimentary level, the primitive cybernetic system discussed in chapter 4, which provides the correct answer to the simple question about the internal temperature of the oven, requires a form of intelligence. Indeed, all cybernetic systems are intelligent to the extent that they must give the correct answer to at least one question. If Gaia exists, then she is without doubt intelligent in this limited sense at the least.

There is a spectrum of intelligence ranging from the most rudimentary, as in the foregoing example, to our own conscious and unconscious thoughts during the solving of a difficult problem. We saw something of the complexity of our own body-temperature regulatory system in chapter 4, although we were mainly concerned with that part which is wholly automatic and does not involve conscious action. Compared with the thermostasis of a kitchen oven, the body's automatic temperature-regulating system is intelligent to the point of genius, but it is still below the level of consciousness. It is to be compared in intelligence with the level of the regulatory mechanisms which we would expect to find Gaia using.

With creatures who possess the capacity of conscious thought and awareness, and no one as yet knows at what level of brain development this state exists, there is the additional possibility of cognitive anticipation. A tree prepares for winter by shedding its leaves and by modifying its internal chemistry to avoid damage from frost. This is all done automatically, drawing on a store of information handed down in the tree's genetic set of instructions. We on the other hand may buy warm clothes in preparation for a journey to New Zealand in July. In this we use a store of information gathered by our species as a collective unit and which is available to us all at the conscious level. So far as is known, we are the only creatures on this planet with the capacity to gather and store information and use it in this complex way. If we are a part of Gaia it becomes interesting to ask: 'To what extent is our collective intelligence also a part of Gaia? Do we as a species constitute a Gaian nervous system and a brain which can consciously anticipate environmental changes?'

Whether we like it or not, we are already beginning to function in this way. Consider, for example, one of those mini-planets, like Icarus, a mile or so in diameter and with an irregular orbit intersecting that of the Earth. Some day the astronomers may warn us that one of these is on a collision course with the Earth and that impact will occur within, say, a few weeks' time. The potential damage from such a collision could be severe, even for Gaia. This kind of accident has probably happened to the Earth in the past and caused major devastation. With our present technology, it is just possible that we could save ourselves and our planet from disaster. There is no doubt of our capacity to send things through space over vast distances and to exercise remote control, with near-miraculous precision, of their movements. It has been calculated that by using some of our store of hydrogen bombs and large rocket vehicles to carry them, we have the capacity to deflect a planetoid like Icarus sufficiently to convert a direct hit into a near miss. If this seems like fantastic science fiction, we should remember that, in our lifetime, yesterday's science fiction has almost daily become factual history.

It might equally well happen that advances in climatology revealed the approach of a particularly severe glacial epoch. We saw in chapter 2 that although another Ice Age might be a disaster for us, it would be a relatively minor affair for Gaia. However, if we accept our role as an integral part of Gaia, our discomfort is hers and the threat of glaciation is shared as a common danger. One possible course of action within our industrial capacity would be the manufacture and release to the atmosphere of a large quantity of chlorofluorocarbons. When these controversial substances, now present in the air at one-tenth of a part per thousand million, are increased in concentration to several parts per thousand million, they would serve, like carbon dioxide, as greenhouse gases preventing the escape of heat from the Earth to space. Their presence might entirely reverse the onset of a glaciation, or at least greatly diminish its severity. That they might incidentally cause some damage to the ozone layer for a time would seem a trivial problem by comparison.

These are just two examples of possible large-scale emergencies for Gaia which we might in the future be able to help her resolve. Still more important is the implication that the evolution of *homo sapiens*, with his technological inventiveness and his increasingly subtle communications network, has vastly increased Gaia's range of perception. She is now through us awake and aware of herself. She has seen the reflection of her fair face through the eyes of astronauts and the television cameras of orbiting spacecraft. Our sensations of wonder and pleasure, our capacity for conscious thought and speculation, our restless curiosity and drive are hers to share. This new interrelationship of Gaia with man is by no means fully established; we are not yet a truly collective species, corralled and tamed as an integral part of the biosphere, as we are as individual creatures. It may be that the destiny of mankind is to become tamed, so that the fierce, destructive, and greedy forces of tribalism and nationalism are fused into a compulsive urge to belong to the commonwealth of all creatures which constitutes Gaia. It might seem to be a surrender, but I suspect that the rewards, in the form of an increased sense of well-being and fulfilment, in knowing ourselves to be a dynamic part of a far greater entity, would be worth the loss of tribal freedom.

Perhaps we are not the first species destined to fulfil such a role, nor possibly the last. Another candidate could be found among the great sea mammals, which have brains many times larger than ours. It is a commonplace of biology that function-less tissues reduce during the course of evolution. Passenger organs do not exist in self-optimizing systems. It therefore seems probable that the sperm whale makes intelligent use of the vast brain it possesses, perhaps at thought levels well beyond our understanding. Of course it is possible that the whale's brain arose for some relatively trivial reason, for example as a multi-dimensional living map of the oceans. Certainly there is no more potent way of consuming memory space than the storage of data in multi-dimensional arrays. Or should we perhaps compare the whale's brain to the peacock's tail, a scintillating mental display organ for the purpose of attracting a mate and enhancing the pleasures of courtship: the whale who provides the most stimulating entertainment having the best choice of mates? Whatever the true explana-tion and however it came about, the real point about the whale and the size of its brain is that large brains are almost certainly versatile. The original cause of their development may be specific, but once they are in existence other pos-sibilities inevitably become exploited. Human brains, for example, did not develop as a result of the natural selective advantage of passing examinations, nor indeed so that we could perform any of the feats of memory and other mental exercises now explicitly required for 'education'.

As a collective species with the capacity to store and pro-cess information, we have probably long surpassed the whale. We are, however, inclined to forget that very few of us as individuals could make an iron bar from iron ore, and still fewer of us could use the bars to make a bicycle. The whale as an individual entity may possess a capacity for thought at levels of intricacy far beyond our comprehension, and might even include among his mental inventions the complete speci-fication of a bicycle; but denied the tools, the craft, and the permanent store of know-how, the whale is not free to turn such thought into hardware.

Although it is unwise to draw analogies between animal brains and computers, it is often tempting to do so. Let us succumb to this temptation and indulge the thought that we humans differ from all other animal species in the superabundance of accessories through which we can communicate and express our intelligence, both individually and collectively, and so use it to produce hardware and to modify the environment. Our brains can be likened to medium-size computers which are directly linked to one another and to memory banks, as well as to an almost unlimited array of sensors, peripheral devices, and other machines. By contrast, whale brains are like a group of large computers loosely linked to one another but almost bereft of any means of external communication.

What should we have thought of an early race of hunters who developed a taste for horsemeat and then proceeded to eliminate the horse from the Earth by systematically hunting and killing every one, merely to satisfy their appetite? Savage, lazy, stupid, selfish, and cruel are some of the epithets that come to mind; and what a waste to fail to recognize the possibility of the working partnership between horse and man! It is bad enough to cull or farm the whale so as to provide a constant supply of those products which whale-hunting nations claim are needed by their backward and primitive industries. If we hunt them heedlessly to extinction it must surely be a form of genocide, and will be an indictment of the indolent and hidebound national bureaucracies, Marxist and capitalist alike, which have neither the heart to feel nor the sense to comprehend the magnitude of the crime. Yet perhaps it is not too late for them to see the error of their ways. Perhaps one day the children we shall share with Gaia will peacefully co-operate with the great mammals of the ocean and use whale power to travel faster and faster in the mind, as horse power once carried us over the ground.

Definitions and explanations of terms

Abiological

Literally without life, but in practice a specialist adjective to describe situations where life has played no part in the end result or product. A piece of rock from anywhere on the surface of the moon has been shaped and formed abiologically, whereas almost all rocks from the Earth's surface have to some extent, great or small, been changed by the presence of life.

Acidity and pH

In common scientific usage acids are substances which readily donate positively charged hydrogen atoms, or protons, as chemists call them. The strength of a solution of an acid in water is conveniently expressed in terms of the concentration of protons it bears. This usually varies from about 0.1 per cent with very strong acids to one part in a thousand million for a very weak acid such as carbonic acid, the acid of 'soda water'. Strangely, chemists express acidity backwards in logarithmic units called pH, so that a strong acid is pH1 and a very weak one pH7.

Aerobic and Anaerobic

Literally with and without air. Words used by biologists to describe environments which are respectively surfeited with or deficient in oxygen. All surfaces in contact with the air are aerobic, as are most of the oceans, rivers, and lakes which bear oxygen in solution. Muds and soil and the guts of animals are all greatly deficient in oxygen and hence called anaerobic. Here live micro-organisms similar to those which inhabited the Earth's surface before oxygen entered the atmosphere.

Equilibrium and Steady State

These technical terms refer to two common conditions of stability. A table stands secure on its four legs and is at equilibrium. A horse standing still is in the steady state because it actively, although unconsciously, maintains its position. If it dies, it collapses.

Gaia Hypothesis

This postulates that the physical and chemical condition of the surface of the Earth, of the atmosphere, and of the oceans has been and is actively made fit and comfortable by the presence of life itself. This is in contrast to the conventional wisdom which held that life adapted to the planetary conditions as it and they evolved their separate ways.

Homoeostasis

A word invented by the American physiologist, Walter Cannon. It refers to that remarkable state of constancy in which living things hold themselves when their environment is changing.

Life

A common state of matter found at the Earth's surface and throughout its oceans. It is composed of intricate combinations of the common elements hydrogen, carbon, oxygen, nitrogen, sulphur, and phosphorus with many other elements in trace quantities. Most forms of life can instantly be recognized without prior experience and are frequently edible. The state of life, however, has so far resisted all attempts at a formal physical definition.

Molarity/Molar solution

Chemists prefer to express the strength of solutions in what they call molarity because this provides a fixed standard for comparison. A mole, or gram-molecule, is the molecular weight of a substance expressed in grams. A molar solution has a concentration of 1 mole of solute per litre. Thus 0.8 molar solution of common salt, sodium chloride, contains the same number of molecules as 0.8 molar solution of an uncommon salt, lithium perchlorate, but because sodium chloride has a lower molecular weight than lithium perchlorate, the one solution contains 4.7 per cent solids by weight and the other 10.3 per cent solids—yet both contain the same number of molecules and have the same salinity.

Oxidation and reduction

Chemists refer to substances and elements which are deficient in negatively charged electrons as oxidizers. These include oxygen, chlorine, nitrates, and many others. Substances rich in electrons, such as hydrogen, most fuels, and metals, are called reducing. The oxidizers and reducers usually react, producing heat, and the process is called oxidation. The ashes and spent gases of the fire can be worked upon chemically to restore the original elements. When this is done to carbon dioxide to make carbon, the process is called reduction. It happens all the time in green plants and algae when the sun shines upon them.

Ozone

A very poisonous and also explosive blue gas. It is a rare form of oxygen in which three instead of two oxygen atoms are joined together. It is present in the air we breathe, normally at only one-thirtieth of a part per million, but in the stratosphere at five parts per million.

Stratosphere

A part of the air lying directly above the troposphere and bounded by the tropopause at 7 to 10 miles high and the mesopause at around 40 miles high. These bounds vary in height with place and season and mark the limits within which the temperature rises rather than falls with increasing altitude. The stratosphere is the site of the ozone layer.

Troposphere

The principal part, 90 per cent, of the air, lying between the Earth's surface and the boundary layer, the tropopause, 7 to 10 miles above. It is the only region of the atmosphere encountered by living things and the place where the weather, as we know it, takes place.

Systems of units and measurement

Many of us are obliged to live in a binumerate state as the old natural system of measurements based upon feet and thumbs and resident in duodecimal or even heptadecimal numbers dies away. Decimal metric scientific units seem so rational and sensible but I suspect that many have more than a sneaking preference for the yard, which can be paced, as against the metre which means nothing real to them. It has even been said that the metric system was part of Napoleon's psychological warfare—a sort of intellectual terrorism to dismay the enemy. The battle between the systems is still on, even after 150 years, and those who imagine that the old system is just some quaint British anachronism should consider that the USA still lives by feet and pounds and gallons and that probably more than half of all the engineering and high technology of the world is in non-metric units. Bearing this in mind, I have used in the text whichever system seemed more appropriate in the context.

Thus to talk of the environmental temperatures in degrees Celsius is less comprehensible to most English-speaking people than to talk in degrees Fahrenheit. Yet no one would list the sun's surface temperature as anything other than 5,500 degrees Celsius or think of boiling liquid nitrogen as other than −180 degrees Celsius.

The convenient prefixes kilo, mega, giga (one thousand, one million, one thousand million respectively) are used to multiply such units as tons, years, and so on. For small quantities the similar prefixes milli, micro, and nano can be used to denote one thousandth, one millionth, and one thousand millionth respectively. Scientific notation is normally used: i.e. 1,500 million is expressed as 1.5×10^9 and one three hundred millionth as 3.3×10^{-9}.

Further reading

CHAPTER 1

Thomas D. Brock, *Biology of Microorganisms*. Prentice-Hall, New Jersey, 2nd edn. 1974.

Fred Hoyle, *Astronomy and Cosmology*. W. H. Freeman, San Francisco, 1975.

Lynn Margulis, *Evolution of Cells*. Harvard University Press, 1978.

I. G. Gass, P. J. Smith and R. C. L. Wilson (eds.), *Understanding the Earth*. The Artemis Press, Sussex, 1971.

CHAPTER 2

A. Lee McAlester, *The History of Life*. Prentice-Hall, N. J., 2nd edn. 1977.

J. C. G. Walker, *Earth History*. Scientific American Books, N.Y., 1978.

CHAPTER 3

B. H. Svensson and R. Söderlund, 'Nitrogen, Phosphorus and Sulphur: Global Cycles', *Scope Ecological Bulletin*, No. 22, 1976.

A. J. Watson, 'Consequences for the biosphere of grassland and forest fires'. Reading University thesis, 1978.

CHAPTER 4

J. Klir and M. Valach, *Cybernetic Modelling*. Iliffe Books, London, 1967.

Douglas S. Riggs, *Control Theory and Physiological Feedback Mechanisms*. Williams & Wilkins, Baltimore, Md.; new edn. Krieger, N.Y., 1976.

CHAPTER 5

Richard M. Goody and James C. Walker, *Atmospheres*. Prentice-Hall (Foundations of Earth Science Series), N.J., 1972.

W. Seiler (ed.), 'The Influence of the Biosphere on the Atmosphere', *Pageoph [Pure and Applied Geophysics]*. Birkhäuser Verlag, Basle, 1978.

CHAPTER 6

G. E. Hutchinson, *A Treatise on Limnology*, 2 vols. Wiley, N.Y. (vol. 1 1957, new edn. 1975; vol. 2 1967).

Robert M. Garrels and Fred T. Mackenzie, *Evolution of Sedimentary Rocks*. W. W. Norton, N.Y., 1971.

Wallace S. Broecker, *Chemical Oceanography*. Harcourt Brace Jovanovich, N.Y., 1974.

CHAPTER 7

Rachel Carson, *Silent Spring*. Houghton Mifflin, Boston, 1962; Hamish Hamilton, London, 1963.

K. Mellanby, *Pesticides and Pollution*. Collins (New Naturalist Series), London, 1970.

National Academy of Sciences, *Halocarbons: Effects on Stratospheric Ozone*. NAS, Washington, D.C., 1976.

CHAPTER 8

R. H. Whittaker, *Communities and Ecosystems*. Collier-Macmillan, N.Y., 2nd edn. 1975.

E. O. Wilson, *Sociobiology: The New Synthesis*. Harvard University Press, 1975.

CHAPTER 9

Lewis Thomas, *Lives of a Cell: Notes of a Biology Watcher*. Viking Press, N.Y., 1974; Bantam Books, N.Y., 1975.

SCIENTIFIC PAPERS ABOUT THE GAIA HYPOTHESIS

J. E. Lovelock, 'Gaia as seen through the atmosphere', *Atmospheric Environment*, **6,** 579 (1972).

J. E. Lovelock and Lynn Margulis, 'Atmospheric homoeostasis by and for the biosphere: the Gaia hypothesis', *Tellus*, **26,** 2 (1973).

Lynn Margulis and J. E. Lovelock, 'Biological modulation of the Earth's atmosphere', *Icarus*, **21,** 471 (1974).

J. E. Lovelock and S. R. Epton, 'The Quest for Gaia', *New Scientist*, 6 Feb. 1975.

'Thermodynamics and the recognition of alien biospheres', *Proceedings of the Royal Society of London*, **B,** 189, 30 (1975).

OTHER RELEVANT PAPERS

I. Prigogine, 'Irreversibility as a symmetry-breaking process', *Nature*, **246,** 67 (1973).

L. G. Sillen, 'Regulation of O_2, N_2 and CO_2 in the atmosphere: thoughts of a laboratory chemist', *Tellus*, **18,** 198 (1968).

E. J. Conway, 'The geochemical evolution of the ocean', *Proceedings of the Royal Irish Academy*, **B48,** 119 (1942).

C. E. Junge, M. Schidlowski, R. Eichmann, and H. Pietrek, 'Model calculations for the terrestrial carbon cycle', *Journal of Geophysical Research*, **80**, 4542 (1975).

Robert M. Garrels, Abraham Lerman, and Fred T. Mackenzie, 'Controls of atmospheric O_2 and CO_2 past, present and future', *American Scientist*, **64,** 306 (1976).

René Dubos, 'Symbiosis between Earth and Humankind', *Science*, **193,** 459 (1976).

Ann Sellers and A. J. Meadows, 'Long-term variations in the albedo and surface temperature of the Earth', *Nature*, **254,** 44 (1975).